1

Mechanical System Interfacing: Introduction

Mechanical systems have become increasingly dependent on computers and electronics to achieve the degree of function, flexibility, and reliability demanded by users. The combination of mechanics and electronics has been called mechatronics, a term that has come to include computers as well. A mechanical system can be characterized as

system = mechanics + electronics + software

The mechanics includes the mechanisms, electromagnetic components, thermal components, flow elements, and so on. The most notable function of the system is usually realized through the mechanics section. The instruments and actuators are considered part of the system mechanics.

The overall control of the mechanical system is expressed in the software. It is the unique role of software that differentiates modern mechanical systems from their forebears. Control of mechanical system through software affords a degree of flexibility unknown to the traditional machine designer. Changing, for example, from manufactured parts of one dimension to parts with another dimension is only a matter of changing numbers in a program. In traditional machines it might have required changing gears, resetting linkages, or other mechanical operations that could take many hours. In addition to this flexibility in usage, such a machine will have fewer moving parts, improving its reliability and probably decreasing its cost.

With this perspective, a new definition for mechatronics can be formulated:

Mechatronics is the application of complex decision making to the operation of physical systems.

This definition recognizes the central role played by software and computers, the decision makers in the design and implementation of mechanical systems. It also recognizes the possibilities of using other technologies for decision making in the future.

Electronics forms the glue that connects the mechanics to the software. At the time the term *mechatronics* was coined, by combining mechanics and electronics, the electronic components themselves were so much more flexible and easy to program than mechanical components that a new class of machine was defined. Software has, however, eclipsed electronics in flexibility and decision making, defining yet another level of intelligent machines. Electronics continues to play a crucial role. That is the role of interface between mechanical components and the computers that run the system software. This role, that of *mechanical system interface*, is the subject of exploration in this book.

1.1 THE INTERFACE

At the lowest level, the interface requires conversion from one energy medium to another and from one power level to another. Because computers are internally electronic, the conversion must have electrical energy as one of its modes. Many conversions require electronics closely coupled to the conversion device, instrument or actuator. Slightly decoupled from the actual energy conversion, electronics provides signal conditioning. This can include change of level for device compatibility, spectral filtering to remove unwanted information, and modulation or demodulation to convert the signal to a more desirable form.

Although software is usually cheaper than custom electronics to design and implement, it has severe speed constraints. Many aspects of processing needed in mechanical systems are simply too fast for software. Electronics can take over data processing chores when software is too slow. These processing chores are usually relatively simple, but crucial. Electronic processing can be either digital or analog, covering such functions as counting (a very common operation), integration, differentiation, generalized logic, selection, and so on.

The ability to build electronics and computing capability directly into a system is leading to new classes of instruments and actuators, *smart instruments*. For example, the traditional boundaries between ac and dc motors are breaking down as the ability to create excitation signals of arbitrary complexity increases. Motors in this category can be made to have the simplicity of dc motors as viewed externally, but actually use ac motor technology internally for a motor with fewer mechanical parts.

1.2 MECHANICAL SYSTEM DESIGN PHILOSOPHY

Mechanical system design has the goals of producing systems with superior performance while minimizing time to market and manufacturing cost, and maximizing reliability.

This suggests the following priority order for choice of medium at the design level:

- Software
- Electronics
- Mechanics

Assigning computational functions to the mechanics, for example, leads to machines with more moving parts and therefore generally lower reliability and higher manufacturing costs. Traditional cam grinding machines, for example, use a master cam and a follower mechanism. Changeover to different cam profiles requires replacement of the master cam and readjustment of the mechanism. Masterless cam grinders use a software (data table) representation of the cam profile to be ground. There is no follower mechanism, and changeover to a new cam profile requires only a new table. New design problems arise as well, however. These involve measurements of tool position and extremely accurate control of the motion of the grinding mechanism to synchronize it with the rotation of the cam blank.

Another example of this is the increasing popularity of brushless dc motors for precision motion applications. The essence of dc motor operation is in the change of electrical excitation to the windings to match the angular position of the motor's rotor. In the Faraday dc motor, this is done mechanically with a set of electrically conducting brushes sliding on an electrically split ring. This commutator changes the excitation every time the brush moves to a new section of the ring. Brushless motors replace mechanical commutation with electrical commutation, improving reliability, particularly in the brushes, which can wear rapidly. But to accomplish the same function, explicit measurement of the rotor angular position is required along with means to change the electrical excitation on command.

Another example arises in controllers themselves. These devices are used to control process or motion variables (pressure, flow, temperature, velocity, position). Historically, mechanics were used for all aspects of the control, measurement, computation, and actuation, as in the Watt governor for steam engines. Controller technology has evolved from mechanical to use of hydraulic or pneumatics, then electronics, and now software.

The operator interface is of critical importance to successful system function and it, too, has changed in response to this design philosophy. Fixed meters, pushbuttons, thumbwheels, and so on, are being replaced with computer display screens, touch-sensitive screen input, keyboards, mice, and trackballs. These computer-based operator interfaces have the potential of providing vastly more effective operator interfaces because information can be grouped according to the immediate needs of the operator, allowing a focus that was not possible in a system with hundreds of fixed elements. On the other hand, the design of the interface becomes even more critical since the operator must have the correct information and controls available to make the right decision.

Overall, this design philosophy can produce systems that:

- Have more functionality
- Are easier to change
- Have a faster development cycle
- Are more reliable
- Give better information and control to the user

However,

- There are new kinds of design problems.
- Different skills are required.
- Unexpected failure modes exist.
- Testing can be more difficult.

1.3 MECHANICAL SYSTEM TIME SCALES

The need for customized electronics in mechanical systems is governed primarily by two factors:

- Signal level
- Time scale

Signal level is connected directly to the conversion process. Standard components, analog-to-digital converters, and other such devices can be used where the signal has a high enough power level to be relatively immune to local electrical noise contamination. Otherwise, electronic signal conditioning is required before these standard devices can be used.

Time scales control the speed with which decisions must be made and actions taken, and thus the speed with which computations must be carried out. The crossover between electronics and software occurs somewhere in the range 1 to 100 μs. Any decisions required below the low end of that range will usually require dedicated electronics. Above the high end, software will generally be used.

A number of time scales are important. Typical time scales for primary actions in representative mechanical systems are:

- *Process systems*: seconds to minutes
- *Large mechanical systems*: tenths of seconds
- *Medium mechanical systems*: milliseconds
- *Small mechanical systems*: 10 μs

The instrument and actuator decision rates can be many times faster than the primary rate. For example, consider the incremental encoder, a common means of measuring linear or rotary position in mechanical systems. It generates a change of state in a digital signal for each specified increment in motion. Even large mechanical systems can generate these changes at maximum rates, in the million per second range. Thus the most basic level of processing must be able to keep up with these rates even though the higher-level decision making can be much slower.

It is interesting that the data rates are often determined by precision/dynamic range considerations rather than raw speed or size of the device being controlled. Precision refers to the size of the smallest change that must be recognized; *dynamic range* refers to the maximum number of such changes that a system must be able to undergo. Precision in such systems is often determined by the need to characterize velocity for low-speed operation. Dynamic range is then determined from the maximum distance the device has to move. The maximum data rate is then dependent on the fastest speed at which the device has to move. One reason for the importance of incremental encoders, for example, is that as digital devices, their dynamic range is limited primarily by the processing equipment rather than by the device itself.

1.4 MAJOR TOPICS

Mechanical system interfacing thus centers on electronics but includes those devices closest to the computational part of the system, primarily instruments and actuators. The book starts with digital logic. With computers playing the central role in mechanical systems, there is a strong motivation in system and component design to use as much digital electronics as possible to simplify the interface. As computers get smaller and cheaper, they can be integrated directly with instruments and actuators. The exploration thus follows naturally from digital logic to this most important application of digital logic, computer architecture. Microcontrollers are designed specifically for such integration and are a hybrid—programmed like computers but used as circuit elements. Motors, the next topics, are the most common actuation elements in mechanical systems. Despite inroads from direct digital instruments and actuators, analog signals are still widely used but must be converted to digital form for computational use. Even in process systems, mechanical motion measurement plays such an important role for such tasks as valve positioning that position and velocity measurement are treated specifically. The final chapters deal with analog electronics, first at the signal level using operational amplifiers and then at the power level.

- Boolean logic
- Sequential logic
- Computer architecture
- Microcontrollers
- Stepping motors
- dc motors
- Analog-to-digital conversion
- Position and velocity measurement
- Analog signal processing (operational amplifiers)
- Power amplification

1.5 PROBLEMS AND DISCUSSION TOPICS

1. Automobile cruise control devices relieve the driver of the need to monitor and control speed. Refer to Figure 1.1 and discuss the following issues:
   (a) Where in the actual vehicle are the signals indicated in the block diagram?
   (b) Identify instrumentation, computation, actuation, and the target system.
   (c) Indicate the power flows in the system and how they are regulated.
   (d) Which aspects of the system are possible candidates for hardware or software control?

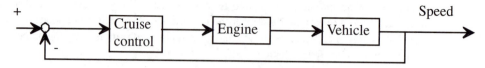

Figure 1.1 Automoblie Cruise Control

2. For the systems listed in (a) to (i) below, identify the:
    - Time scales
    - Degree of flexibility
    - Computational medium (software/hardware and what kind)
    - Instruments used
    - Actuators
    - Market size (number of units sold)
    - Operator interface

Sample systems:

(a) Temperature regulation in a petrochemical plant
(b) Temperature regulation in a private home
(c) Temperature regulation in a commercial building
(d) Motion of axes in a machine tool
(e) Automatic coffeemaker for home use
(f) Microwave oven for home use
(g) Motion of the read-write head in a disk drive
(h) Control of beam scanning in a television
(i) Attitude control in a communications satellite

3. Consider the following two applications for two-axis (XY) tables:

    (i) *The XY axes of a milling machine:* 1 m of motion in each direction, capable of being used to cut steel, 0.02 mm precision

    (ii) *An XY table used for inspection of integrated circuit wafers after manufacture:* 20 cm of motion in each direction, microscope/TV inspection, 0.3 μm precision

    Discuss the various actuation options listed below in light of issues such as cost, additional equipment (pumps, compressors, etc.), maintenance requirements, smoothness of motion, available force, range of available speeds, and ease of doing closed-loop control.

    (a) Pneumatic cylinder
    (b) Hydraulic cylinder
    (c) Ballscrew and rotary motor
    (d) Chain or belt drive and rotary motor
    (e) Rack-and-pinion drive with rotary motor
    (f) Direct-drive linear motor

4. Using a computer equipped with a lab I/O interface that can read or write logic signals, determine the maximum speed at which:

    (a) A single output signal can be turned on and off in succession
    (b) An input signal can be read and copied to an output
    (c) Two input signals can be read, inverted (0 becomes 1; 1 becomes 0), then sent out on two output channels
    (d) A sequence of input pulses can be counted without losing any counts

2

Combinational Digital Logic

2.1 QUANTIZED SIGNALS

Information in a signal is established by a coding convention. The simplest coding associates the information with one of the signal's primary physical variables: in the case of electrical signals, either voltage or current. Once that association is made, though, a convention delineating how much information the signal carries, that is, how many signal levels can be distinguished, must also be decided. In analog signals, there is no limit to the information content by convention—the limitations arise from the resolving power of the receiving instrument and the noise present with the signal. In this context, *information* refers to the number of individual values that can be resolved from a signal. If the information is associated with a voltage, for example, as is most common, an allowable voltage range can be represented as a real number. Since there are an infinite number of real numbers on any finite section of the number line, there is no theoretical limit to the information content. It is also possible to treat the voltage as part of the integer number line, so that in any finite section there are only a finite number of integers. When this form of coding is used, the signal is said to be *quantized*. If the quantization interval is larger than the noise present, the noise will have very little effect on the interpretation of the signal, so the actual information content will match the theoretical content very closely.

Binary is the ultimate quantization—the signal is divided into only two values. Although this may seem drastic and uneconomical, allowing for only two decision states when technology to get hundreds or thousands exists, it has turned out to be crucially important in computer system design. By using only two states, systems can be built that have a negligibly small error rate, so small that for engineering purposes they can be viewed as being completely predictable regardless of their complexity. This property has been exploited to design computers and other logic systems of far greater complexity than that ever dreamed possible by those who proposed that computers be built using binary logic elements.

2.2 ERROR-FREE CODING

Just quantizing a signal does not give the nearly error-free operation needed for large-scale digital circuits. If a signal is near the boundary, a small amount of noise could cause its value to change. A buffer zone provides some protection against this possibility. The two values are separated by a distance large enough so that noise cannot move one value to another. But what does it mean if a value is in the buffer zone? Small noise disturbances could push a signal into the buffer zone, where it would be undefined. It would tend to stay close to the zone from which it was "pushed" and would thus be likely to be classified as such by a receiver. "Likely" isn't good enough, however. Statistical noise rejection cannot provide enough error protection for high-speed digital circuits, which have many components and operate at high repeat rates. To give even better protection, component specifications are written so that the allowable voltage ranges are different for input and output. Since outputs are always connected to inputs, the idea is to establish an input buffer zone that completely encompasses the output buffer zone. Thus a small amount of noise on an output signal cannot move it out of the buffer zone recognized by the input.

For example, the TTL (transistor–transistor logic) specification, which has a nominal value of 0 volts for one of the binary values and 5 V for the other, uses the following:

Output:

  • For binary value 0, output voltage < 0.4 V
  • For binary value 1, output voltage > 2.4 V

Input:

  • For binary value 0, input voltage < 0.8 V
  • For binary value 1, input voltage > 2 V

This gives an absolute protection zone of 0.4 V. As long as the noise is less than 0.4 V, a 0 and a 1 can never be confused. As an example, if an output signal is a logic 0 and has a voltage of 0.39 V, an additive noise of 0.1 V could make the signal appearing at the input of the next component have a value of 0.49 V. For inputs, however, this is well within the "0" range of 0 to 0.8 V.

2.3 STATIC/DYNAMIC, ALGEBRA/CALCULUS

Binary digital systems can be made nearly error free, and thus totally predictable, but the very low information content of each signal means that even relatively simple problems can result in complex circuits. Design methods are needed that can produce reliable circuits from engineering descriptions of the desired behavior, despite the complexity. Design of logic systems can be broken into two phases: static and dynamic. These follow the same breakdown as that used in "ordinary" systems:

> *Static systems* have outputs that depend only on the current values of the inputs. They can be described by algebraic equations.

> *Dynamic systems* have outputs that depend on past values of the inputs as well as the current values. They require calculus to describe them mathematically.

In digital logic (the term *binary* will be dropped from this point forward; it will be assumed that digital systems are binary unless otherwise noted) these definitions also apply. There is an algebra, Boolean algebra, based on the work of an English mathematician, George Boole (1815–1864), but there is no formal calculus. The methods of sequential logic are used to design dynamic digital systems by reducing dynamic problems to the solution of a set of static problems. The design methods translate the problem description to a mathematically rigorous form, generate solutions as logical expressions, provide means for optimizing solutions to eliminate redundant elements and for testing for intrinsic problems (hazards), and finally, couple directly to implementation means. This seems like a tall order, and it is. It usually works quite well, a tribute to the leap-of-faith commitment to binary quantization.

2.4 COUNTING AND BINARY NUMBERS

Boolean algebra uses variables that can have only two values, 0 or 1. These variables are often used in the design of systems that, internally, must represent ordinary numbers; the computer itself is the most common example. To do that, large numbers of Boolean variables are combined to produce groups of signals that, together, can represent an ordinary number. Because the Boolean variables have only two values, the natural connection is to base 2 (binary) numbers. Each Boolean variable can thus represent one binary digit. Several binary counting systems are encountered in system design. The positive integers are the best known, 0, 1, 10, 11, 100 for 0, 1, 2, 3, 4, and so on. Negative numbers are most often represented in 2's–complement form. This form is convenient for circuit design but is not immediately intuitive. Two's complement numbers are formed by counting backwards from zero using a finite-precision word. With a three-bit word, for example, the counting backwards from 000 gives 111; similarly, 111 + 1 = 0, as long as the carry is thrown away. Thus 111 represents the number -1. Subtracting 1 again gives 110, which is thus -2. With only a finite number of combinations, there must be a border between the positives and the negatives—all numbers with the left bit equal to 1 are negative.

100	-4		000	0
101	-3		001	1
110	-2		010	2
111	-1		011	3

Note that although the left bit separates positives and negatives, it is *not* a sign bit. All bits must be used to convert properly.

2.5 GRAY CODE

Natural binary, described above, is not the only way to code numbers. It is convenient and mathematically intuitive because it uses the standard place-value number system. But from a circuit point of view, there can be problems associated with the fact that adjoining numbers can have differences in several bits. For example, 000 to 111 has changes in all of its bits; 101 to 100 has changes in two bits, and so on. Gray counting systems do not use place value but, instead, are built on the rule that adjoining numbers always have changes in only a single bit.

As an example of a Gray code application, it is common for two processes to exchange digital data, where the data are represented by more than a single bit. If the two processes are not synchronized with each other, there is no way to tell when the data exchange will actually take place. For two-bit data represented in natural binary, for example, if the data value were to change from 1 (01) to 2 (10), a problem could occur if the receiving process was reading the data while the sending process was sending the data. Since two signals can never be changed exactly simultaneously, one would have to be changed before the other. The problem occurs if the receiving process reads the data in the middle of the change. Depending on which bit were changed first, the intermediate values would come from the sequences

01–11–10 or 01–00–10

In either case, the intermediate value is totally different than either the beginning or ending values; in one case the receiver would get the value 11, in the other 00. However, if instead of natural binary counting order, the following definition of counting order is used:

00 01 11 10

this problem cannot arise. In each case, adjacent values change by only one bit, so regardless of when the value is read, erroneous data will never be transmitted. This counting order is called a Gray code. In this case, the corresponding integer values are

00	0
01	1
11	2
10	3

which appear a bit strange, but certainly represent a valid code. This code is a *reflected* Gray code. To construct code sequences with any number of bits, start with the one-bit sequence

    0 1

which has no simultaneity problems. To construct a two-bit sequence, write the first two numbers by putting a 0 in front of each of the one-bit sequence entries,

    00 01

For the next two numbers, reverse the order of the one-bit sequence for the right bit, and use a 1 for the left bit:

    11 10

or 00 01 11 10 for the full sequence.

To get a three-bit code, do this again:

    000 001 011 010 (two-bit sequence + leading 0)   then
    110 111 101 100 (two-bit sequence in reverse order with left bit set to 1)   for
    000 001 011 010 110 111 101 100

If Gray code is used, the input signal can be read with no synchronizer because the input must go through all intermediate values to get from one value to another. Reading at any time will give a valid result. If the value is changing, the result will be either the new value or the old value—never a spurious value. For non–Gray code inputs, a synchronizer (clock) or handshake must be used.

2.6 BOOLEAN ALGEBRA

Variables in Boolean algebra can take on only two values, for which the symbols 0 and 1 are normally used. These symbols should not be confused with the numerical values that use the same symbols in normal algebra. A variable can be set to a value,

    $x = 0$ $x = 1$

These are the only possible values for x. These Boolean variables do not correspond to computer variables as used in languages such as C or Fortran. Those variables refer to computer words. Since each computer word encompasses many bits, at the circuit level it would be represented by a set of Boolean variables, one variable for each bit.

Boolean algebra has one unary operator, NOT, which is usually represented by a prime on the variable or with a bar over it:

    x' $\overline{x}$

The NOT operator, as with all operators in Boolean algebra, can be defined by exhaustively listing all possible combinations and the associated result. Since x can only take two values, the list (called a truth table) for NOT is very short,

x	x'
0	1
1	0

The two standard binary (i.e., they take two operands) operators are AND and OR. The truth tables for these follow the standard logical definitions.

xy	x and y	x or y
00	0	0
01	0	1
10	0	1
11	1	1

An alternative form for the truth table is as a map. The exclusive-OR (XOR) operator is shown as a map in Figure 2.1. The map format is used extensively in function minimization, introduced later in this chapter.

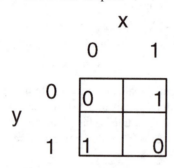

Figure 2.1 Map of Exclusive-OR (XOR)

2.7 BOOLEAN EQUATIONS

Boolean algebra also borrows symbols for its operators from ordinary algebra,

    ∗ (or nothing) for AND
    + for OR
    ´ for NOT (or bar)

Equations are written in the normal way, using parentheses and a hierarchy that gives ∗ precedence over +. For example, in the expression

    x ∗ (y + z) + y´

the parenthetic expression (y + z) is evaluated first. In the expression

    xy + z

the (implied) ∗ is evaluated before the +.

    This hierarchy can be deceptive, however. Unlike multiplication and addition, AND and OR are completely equivalent and, in fact, are dual. Duality in this instance means that a complete inversion of definitions (0 for 1, AND for OR) leaves the results unchanged. Thus the hierarchy is a convenience in writing equations but does not express anything fundamental about Boolean algebra. The duality is often useful in component design, however, since there may be economy or performance reasons to prefer one form over another.

2.8 AXIOMS

Boolean algebra is based on a set of axioms. These unprovable statements establish the ground rules for all operations in Boolean algebra. They deal with the basic properties of the algebra, such as the definitions for values, commutativity, and so on.

- *Commutativity:* $x + y = y + x$; $x * y = y * x$
- *Associativity:* $(x + y) + z = x + (y + z)$; $(x * y) * z = x * (y * z)$
- *Distributivity:* $x + (y * z) = (x + y) * (x + z)$; $x * (y + z) = (x * y) + (x * z)$

The first of the distributivity relations shows the duality in Boolean algebra. The OR (+) operator distributes in the same manner as the AND (∗) so while the second relation looks "normal," the first seems rather strange.

- *Existence of 0,1:* $x + 0 = x$; $x * 1 = x$
- *Existence of complements:* $x + x' = 1$; $x * x' = 0$

The axiom for existence of complements also is strange looking. The complement operator does not have a direct counterpart in ordinary algebra (although rough equivalents to negation and inversion can be drawn).

2.9 THEOREMS

Theorems are statements about Boolean algebra that can be proven using the axioms. There is no limit to how many theorems can be written down. The ones listed here are important to logic system design because they are used in the operations of creating and optimizing combinational logic circuits. They can be proved using algebraic manipulation or by exhaustive substitution. Proof by exhaustive substitution, although more of a brute-force method than algebraic manipulation, is no less rigorous from a mathematical point of view.

- *Idempotent operators:* $x + x = x$; $x * x = x$
- *Absorption:* $x + x * y = x$; $x * (x + y) = x$
- *Simplification:* $x + x' y = x + y$; $x * (x' + y) = x * y$
- *DeMorgan's laws:* $(x + y)' = x' * y'$; $(x * y)' = x' + y'$

2.10 LOGIC SYSTEM DESIGN

The goal of logic system design is a functioning circuit: that is, an electronic system whose outputs match the system specifications for all inputs. Boolean algebra will be used to proceed from the design specification to an implementable logic function. In practice, computer-aided design tools are used for this process for all but the most simple systems, although the manual procedure will be outlined here.

Design problems that result in static systems (i.e., Boolean or combinational) have the following properties:

- Binary inputs
- Binary outputs
- No history dependence

For example, consider a simple electronic door lock with a set of buttons. To enter, a person must press the correct combination of buttons simultaneously. Assume the following logic conventions:

- *Inputs:* button in = 1, button out = 0
- *Output:* lock actuator, 0 lock, 1 unlock

That is, on input, a button pressed in will generate a logic 1; when it is not pressed it will generate a logic 0. On the output, a logic 1 signal is required to unlock the door. The voltages associated with these logic levels depend on the implementation. For example, an implementation could be used that associates a logic 1 with a high voltage (e.g., 5 V for TTL) and logic 0 with ground-level voltage. A truth table for this design of a two-button lock is shown below. $x1$ and $x2$ are the logic variables representing the input pushbuttons, and L is the lock output signal.

x1	x2	L
0	0	0
0	1	1
1	0	0
1	1	0

Even for this two-variable problem, generation of a logic function for this truth table is not simple. What is needed is a systematic procedure for converting the truth table into a set of implementable logic equations.

2.10.1 Function Generation from Truth Tables

The procedure starts with a Boolean function that expresses all possible combinations of the input variables and their primes,

$$y = c1 * x1 * x2 + c2 * x1 * x2' + c3 * x1' * x2 + c4 * x1' * x2'$$

The coefficients $c1, c2, \ldots$, are unknown at this point. This is called the *sum-of-products* form for a logic expression. There is also a dual form, *product of sums*. It is completely equivalent to the product-of-sums form but is used less because it looks less natural.

The coefficients can have logic values of 1 or 0, indicating whether a particular term should be included or not in the final logic expression. This determination is made by comparing the terms first to the input columns of the truth table. For a term in the sum-of-products expression, match the row of the truth table by calling an unprimed variable 1 and a primed variable 0. For example, the term $x1 * x2$ has both variables unprimed, so will match the row of the truth table with 11 for inputs. The term $x1' * x2$ will match the row with 01 as inputs. Once a term is matched, set the coefficient to 1 if the matching row has an output of 1; if the output is 0, set the coefficient to 0 (i.e., omit that term).

To get the Boolean expression corresponding to a truth table, then, only the rows of the truth table that have an output of 1 need be considered, since the rows with 0s will not contribute to the final expression. In this case, only the row for x1 = 0, x2 = 1 has an output of 1. This matches the term x1′ * x2 in the sum-of-products expression, so the logic equation for the lock output is

L = x1′ * x2

All Boolean design procedures flow from this technique. It allows for arbitrary synthesis of logic functions. It is the Boolean equivalent of regression, but it is exact. That is, by listing all the input/output sets, an exact Boolean function can be derived as a functional representation of the data set.

The design procedure can be repeated for the product-of-sums form. The complementary form of the sum of products is

y = (x1 + x2) * (x1′ + x2) * ⋯

The rules are exact duals of the rules to construct the sum of products:

• Each truth table entry that has a 0 output generates an output term.
• Within each term, write variables with 0 inputs unprimed and variables with 1 primed.

For the two-button door lock:

L = (x1 + x2) * (x1′ + x2) * (x1′ + x2′)

Note that this is considerably more complex than the sum-of-products form. This is coincidental; in general, the complexity will depend on the problem specifics. For example, consider the case where L = 0 unlocks the door, which might be done if it were important for safety reasons that the door unlock on a power failure. In this case, the product-of-sums form would generate a simpler expression. To avoid complexity problems of this sort, procedures are needed to simplify logic expressions. This can be done algebraically, but the algebraic procedure is not systematic and can be tedious, so algorithmic and tabular methods are generally used.

2.11 LOGIC FUNCTION MINIMIZATION

The most commonly used manual minimization method is based on the map representation of truth tables (Figure 2.1). Once the map is drawn, repeated application of the theorem

xy + xy′ = x

is used to generate a reduced expression. What this theorem states is that if in two terms one variable appears primed in one term and unprimed in the other, and if the terms are otherwise identical, the two terms can be combined with only the identical part left.

This theorem can be applied graphically by arranging the map in such a way that adjacent cells differ by only one variable, which appear primed in one and unprimed in the other. Then any pair of adjacent cells that both contain a 1 (for sum-of-products form) will generate only a single term in the final expression, and that term will reflect only the unchanged variables. This can be repeated with pairs of pairs, and so on, so that any set of 2^n adjacent cells, all of which have 1s, can be combined to a single term containing only the variables that do not change over the entire set. This form of logic truth table is called a *Karnaugh map*.

Adjacency requires that only one variable change between adjoining cells, which is easy for a two-variable map, which is two-dimensional (Figure 2.2). Imagine a door lock with two possible combinations (perhaps one for building operator), as shown in Figure 2.3. This has two 1s, so the solution from the truth table would be

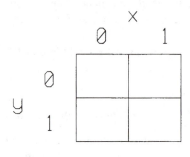

Figure 2.2 Two-Variable Map

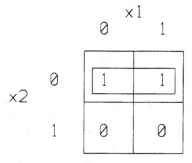

Figure 2.3 Lock with Two Combinations

$L = x1' * x2' + x1 * x2'$

The theorem above shows that this can be reduced to just

$L = x2'$

Reference to the map (Figure 2.3) shows the same thing; the two terms with output equal to 1 are adjacent, and a box is drawn to show that they can be combined to a single term. Reference to the labels shows that $x2'$ is common to both.

If an alternative combination had been chosen, Figure 2.4, there are no adjacent 1-boxes. This map gives the function,

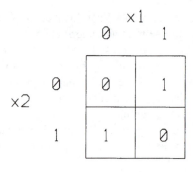

Figure 2.4 Alternative Lock Combination

$$L = x1*x2'+x1'*x2$$

which is not reducible by this theorem. Note that the map method does not guarantee absolute minimization, just minimization for product-of-sum (or dual) form. Such designs may thus include some unnecessary redundancy. Some automated design methods may carry the minimization further.

2.12 KARNAUGH MAP

A three-variable map will be three-dimensional and not directly representable on paper, so a means is needed to map it to two dimensions. This is done by arranging the variables in groups and then ordering them in such a way that adjacency is preserved. The order that is used is that of a Gray code. Because a Gray code guarantees that adjacent numbers always change by only a single bit, they will then be adjacent in the sense needed for application of the reduction theorem.

To construct a truth table in map form for more than two variables, use Gray code counting order for adjacent boxes, as in Figure 2.5 for a map with three variables. Because the Gray counting order is used along the axis where two variables are defined, all of the boxes are adjacent (only one bit, i.e., variable) changes value from box to box. Examination of the map also shows that adjacency is established along the exterior boundaries (i.e., a box can "wrap around."). If the map of Figure 2.5 represented a three-button door lock, the boxed pair of 1s would give the lock function,

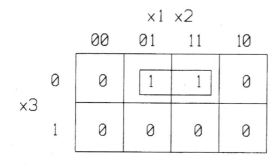

Figure 2.5 Three-Variable Karnaugh Map

$$L = x2*x3'$$

Adjacent pairs of pairs can be eliminated in a single step (Figure 2.6). In this case, both x1 and x3 change inside the box, so the result is

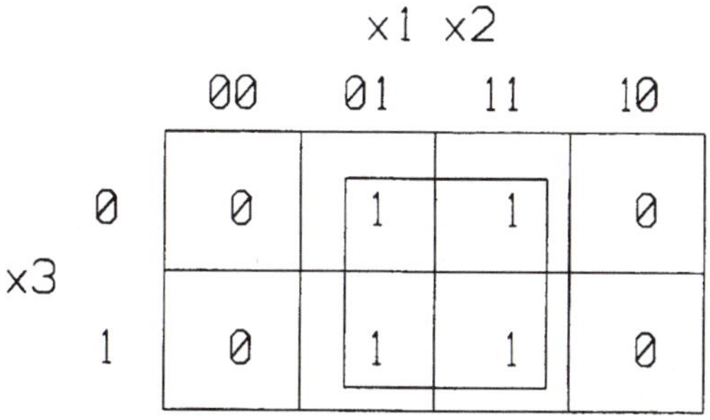

Figure 2.6 Three-Variable Karnaugh Map with Four-Cell Term

L = x2

This can be extended to four variables (Figure 2.7).

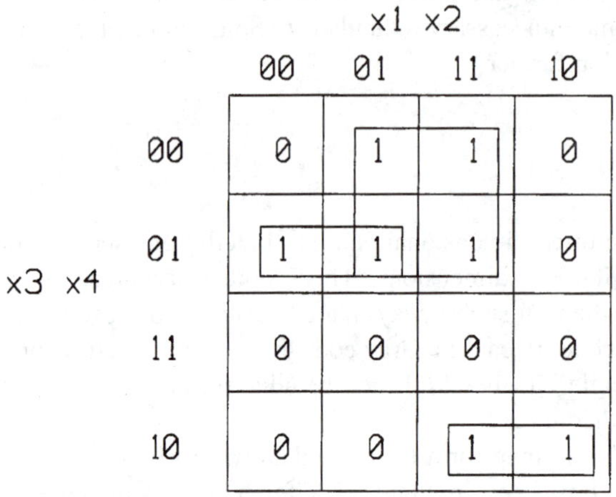

Figure 2.7 Four-Variable Karnaugh Map

This map gives the expression,

L = x1´*x3´*x4+x2*x3´+x1*x3*x4´

This method can be extended all the way to six variables by using multiple tables. Beyond six, other methods for minimization must be used.

2.13 DON'T CARE COMBINATIONS

In many logic systems there are input combinations that are considered "impossible" to occur, or represent input conditions for which there is no specification for the output. These signals are all electrical, however, and there cannot be "no" value of voltage. These conditions can be used either for error checking or for further circuit minimization.

The illegal or impossible input conditions can be used for error checking by adding outputs to a circuit that indicate when a malfunction might be taking place. These outputs are ON when the impossible combinations take place. On the other hand, by carefully examining a system's Karnaugh map, the cells that correspond to the don't-care states can be set to either 0 or 1 in such a way that the size of the design groupings is increased, thus simplifying the output equations.

2.14 HAZARDS

Glitches are short, spurious signals generated when inputs are changing. Consider the map in Figure 2.8. The function for this map is

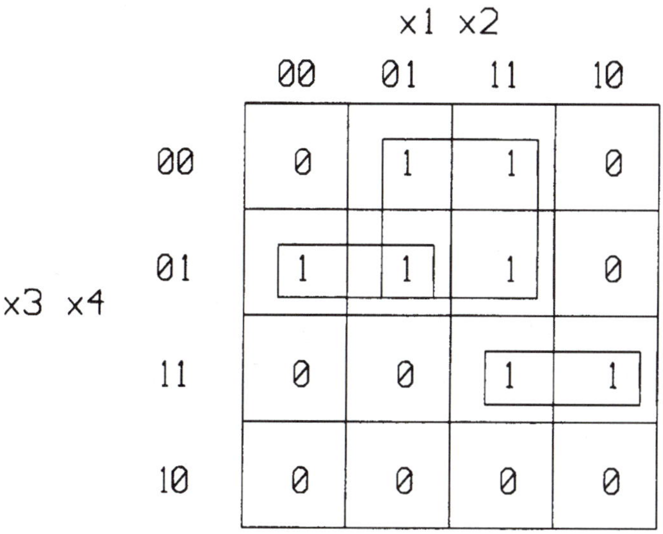

Figure 2.8 Karnaugh Map Showing a Potential Hazard

$$y = x1'*x3'*x4+x2*x3'+x1*x3*x4$$

During an input change from 1101 to 1111, the output should remain at 1. The first term, $x1'*x3'*x4$, is 0 for both inputs. The second term changes from 1 to 0 as the input changes, while the third term changes from 0 to 1. In real systems, however, these changes will not be simultaneous, so the circuit output has a possibility of momentarily becoming 0 during the transition, since the first term is also 0. This condition is referred to as a hazard.

If the device connected to the circuit output is static, that is, its outputs depend only on the current values of the inputs, an input glitch will show up as an output glitch, a momentarily incorrect value. This might not cause serious problems since the glitch is normally very short. Low-pass filtering devices (like motors) will ignore a short glitch entirely. If device is dynamic (sequential), however, it might respond to the momentary change, if the change is long enough to switch one of its inputs. For example, a counter might increment its count in response to a spurious input. There will be no way to correct or otherwise detect the fact that an erroneous count was made. The remedy for this is to "deminimize" the circuit in a controlled way. If for all transitions there is always one term that does not change during the transition, there will be no glitch. This can be done by making sure that no transitions go from one grouping to another without being covered by another grouping. That is, all transitions between adjacent boxes should take place under a comnmon term, which will not change during the transition, thus protecting against spurious outputs.

The Karnaugh map with an added term to guard against hazards is shown in Figure 2.9 on the following page. The protected function now is

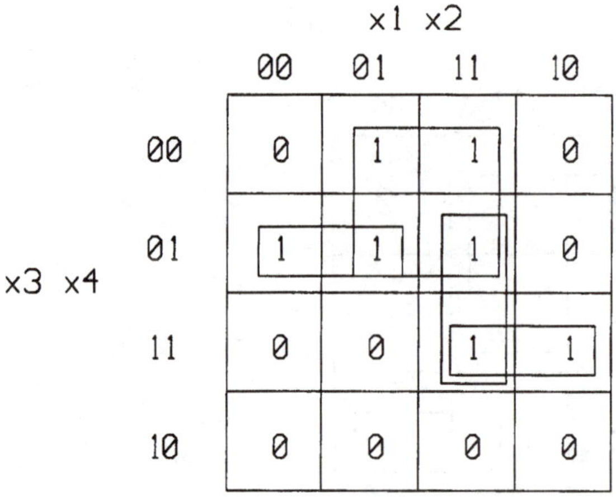

x1 x2

	00	01	11	10
00	0	1	1	0
01	1	1	1	0
11	0	0	1	1
10	0	0	0	0

x3 x4

Figure 2.9 Map with Term Added for Hazard Protection

$$y = \quad x1'*x3'*x4+x2*x3'+x1*x3*x4+x1*x2*x4$$

The last term is 1 throughout the troubling transition. All transitions in this map are now protected against *static hazards*.

2.15 PHYSICAL REALIZATION OF BOOLEAN FUNCTIONS

Combinational logic—static, Boolean—because of digital quantization, is relatively easy to achieve. The circuits designed are almost exact representations of the mathematics. Real circuits have some delay, but that is not usually a serious consequence for combinational systems. It can be very important when the combinational system becomes part of a sequential system.

2.15.1 Relay Logic

Logic systems made from solenoids and contacts (relays) were, at one point, the primary way to implement logic functions. The logic values 0 and 1 are represented by open or closed circuits. Two contacts in series, for example, provide a logical AND, while contacts in parallel give a logic OR. The relays can be normally open or normally closed, to represent inputs of a variable or its complement. Since the output is an open or closed electrical contact, it can be used to activate a power system as the system output.

Relays are no longer used for large-scale logic because they are very slow, large, and expensive compared to electronic logic. They are still used for switching signals from very high impedance devices, such as thermocouples, because the relay contact has much lower impedance than that of equivalent solid-state switches. They are also sometimes used for switching high power. An important heritage of relay logic, however, is the ladder diagram, a method of representing logic systems. It is widely used as the standard programming input for programmable logic controllers.

2.15.2 Programmable Logic Controllers (PLC)

Programmable logic controllers (PLCs) have replaced relays in many industrial systems and replaced hardwired electronic logic in systems where the electronic logic had replaced relays. A PLC is a computer programmed to implement logic functions. Its characteristics are that it has ruggedized construction for industrial use, using isolated inputs and outputs, and is programmed in ladder logic, which is a graphical representation of relay logic. They are popular for implementation of logic control functions because they are easy to use and the ladder language is widely known. They are often used in the control of manufacturing systems.

2.15.3 Implementation Conventions

In implementation it is important to remember that 0, 1 are dual and that there is no inherent mathematical preference for 0 or 1 for connection to physical signal. The following are common defaults, but reverse conventions are also used:

- *Relay*: as above, closed = 1, open = 0.
- Normally open relays can be cascaded without causing complementation.
- Use normally closed relay to generate complement.
- *Electrical:* high voltage = 1, low voltage = 0.

Opposite conventions are often mixed in the same system to increase efficiency or to match default or safety conditions for different components.

2.15.4 Fluid Logic

Fluid logic can be implemented with hydraulics or pneumatics and can be purely fluid or mixed domains, with mixtures of fluid and mechanical parts. Valves act very much like relays and can be used to implement logic functions. Like relays, they are no longer used for large-scale logic functions.

Purely fluid logic elements are called fluidic components. They have no moving parts, so can be viewed as the fluid counterparts of solid-state electronics. They use a variety of fluid phenomena, such as vorticity or wall attachment, to realize logic functions. Fluidics had very high profile when they were developed in the 1960s, but have since become niche products. They are important in electrically adverse environments, such as explosive environments, or in devices such as jet engines with a combination of a ready supply of working fluid and a harsh environment for electronics.

2.15.5 Electronic Logic

Electronics dominates logic implementation technologies. A number of electronic technologies are available; transistor–transistor logic (TTL) has been the most common technology for small scale logic functions, while integrated circuits using a variety of technologies are used for large-scale circuits that will be reproduced in large numbers. CMOS (complementary metal oxide semiconductor) technology has also become very common and has electrical characteristics very close to those of TTL, so they can be interfaced easily. CMOS is desirable because it can have very low power consumption, particularly in the quiescent state, but it is very sensitive to damage from static electricity. Even where TTL is not used internally, as in computers, it is normally the standard convention for connecting inputs and outputs to a computer.

2.15.6 Logic Block Diagrams

The Karnaugh map provides a procedure for taking an engineering description of a logic problem and deriving an algebraic function for it. The next step is to develop a circuit from the algebraic description. This is done most easily by creating a logic block diagram from the algebraic description, then a circuit from the logic block diagram. Logic circuit behavior is so close to ideal that the logic block diagram can be considered to be a pseudo-circuit. The only conversion required is to recognize practical limits of the actual circuit components, such as the maximum number of inputs that can be connected. Component substitutions might also be made for performance purposes: for example, to maximize speed or minimize power consumption.

The logic block diagrams are constructed from the standard components shown in Figure 2.10. The NAND element is a combination of NOT and AND:

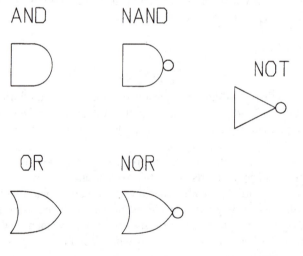

Figure 2.10 Standard Logic Elements

$$\text{NAND}(x,y,...) = (x * y * \cdots)'$$

and the NOR is a combination of NOT and OR:

$$\text{NOR}(x,y,...) = (x + y + \cdots)'$$

These operators are important because that is how many real elements work.

2.15.7 Door Latch in Circuit Form

The earlier solution to the two-button door latch,

$$L = x1' * x2$$

can be converted to pseudocircuit elements (Figure 2.11). If an implementation using NANDs and NORs is needed, this logic diagram can be converted to NAND/NOR form by using DeMorgan's theorems to get the equivalents,

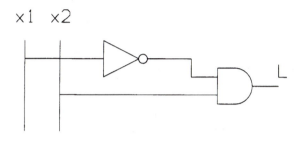

Figure 2.11 Door Latch Circuit

$$(x + y)' = x' * y'$$

$$(x * y)' = x' + y'$$

This then gives equivalents for AND and OR:

$$x * y = (x' + y')'$$

$$x + y = (x' * y')'$$

The logic block equivalents are shown in Figure 2.12. This seems to give lots of inverters, but many of them will cancel out in constructing an actual circuit.

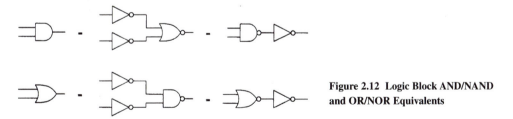

Figure 2.12 Logic Block AND/NAND and OR/NOR Equivalents

2.15.8 Practical Considerations

Actual circuit design depends on a number of practical factors as well. Fan-in and fan-out are the most prominent. These properties describe the number of connections that can be made to a given logic element. *Fan-in* is the number of inputs; *fan-out* describes the number of other elements that can be driven from a logic element's output. This is a current limitation and is normally quoted in terms of the number of standard TTL loads that can be driven; a normal TTL fan-out is 10 other TTL devices of the same type. The limitation on how much current a device can supply also applies when logic devices are connected to other system components, such as LED displays, motor drivers, and so on. In each case the current requirement of the device using the logic signal should be compared to the capability of the logic device to supply current. If there is a mismatch, an interme- diate amplifier may be needed to boost the current supply capability.

As components operate, their need for current varies. These transients can cause component interaction through the power supply circuitry. Capacitors placed at power input to each active component provide local power to meet transient needs. As a general rule, decoupling capacitors for each integrated circuit should be 0.01 to 0.1 µf. Decoupling capacitors also offer a low-impedance path to ground for high-frequency noise signals that might be superimposed on the supply voltage and thus assist in con- taining the spread of electrical noise.

The output transistor of a logic component (gate) acts as a switch which can be con- nected to ground or left open. To convert this to a voltage, a *pull-up resistor* is used. The pull-up resistor is installed from the output to a 5-V source (Figure 2.13). When the switch is open, there is (ideally) no current flow through the pull-up resistor, so the volt- age measured at the output will be the same as the source voltage, 5V. When the switch is closed, the output point will be grounded and the voltage measured at the output will be zero. During this condition, there will be current flow through the pull-up resistor. As long as the resistance of the switch is much lower than that of the pull-up resistance, the out- put voltage will stay very close to zero. There will, however, be heat dissipation associ- ated with the current flow in the pull-up.

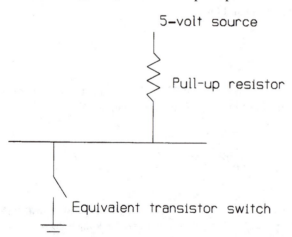

Figure 2.13 Pull-Up Resistor on Output of Logic Gate

If a device does not have an internal pull-up, it is called an *open-collector device*. In that case an external pull-up must be supplied (of order 1 KΩ); small pull-ups are used for high speed but have high power drain and attendant heat generation. In general, there is a trade-off between power and speed in logic devices of the same family. High resis- tance in the pull-up uses less power, but in coupling with the parasitic capacitances in the devices causes longer time constants and thus slower switching time.

2.15.9 Connecting to a Bus

A *bus* is a parallel collection of wires that is often used to interconnect electronics components, particularly computer peripheral components. The idea of a bus is that any of the connected devices can either receive information from the bus or put information on the bus. Receiving is no problem as long as the device putting the information on the bus can produce sufficient current to drive the receiving devices. Putting information on the bus, "writing" to the bus, is another matter, however. Only one device output can be connected for writing; otherwise, the conected devices will "fight" each other and the result will be indeterminate. If multiple devices are connected simultaneously and some are attempting to assert lines high while others are attempting to assert the same lines low, the results will typically be indeterminate.

Open-collector outputs were an early solution for this. By connecting a number of open-collector outputs together, with a single pull-up, there will be no conflict as long as all the switches are open. When one device wants to write to the bus, the result will be high voltage (pulled up) if that device's switch is open, or low voltage (ground) if that device's switch is closed. The rule for nonconflicting operation is that all non-active devices must have their switches open. This connection is also called a *wired-OR*, or *wired-AND*, depending on the logic convention, since it implements the OR or AND function without a gate (i.e., no active component). This is a "free" way of getting a gate.

An alternative to an open collector for driving a bus is the use of *three-state logic*. This is not trinary logic; rather, it involves the use of an actual switch to disconnect inactive inputs from the bus. When disconnected, the device is referred to as being in its *high-impedance state*. This arrangement is more noise immune and can be made faster than the open collector method of bus connection, so is nearly universal now. It is, however, a more complex circuit and requires an additional "enable" input in order to control the three-state switch.

2.15.10 Analog-to-Digital Conversion: The Schmitt Trigger

It is often necessary to convert an analog signal to a digital value. Most commonly this occurs when there is a logic signal whose value depends on whether an analog voltage is less than or greater than a specified value. The comparison is done with an analog component (see Chapter 11) and the result must be converted to a one-bit digital value. The Schmitt trigger is a device that is often used as the last stage of this conversion. The Schmitt trigger uses a hysterisis zone (see Figure 2.14 on next page), so that small oscillations in the analog signal do not cause the logic signal to oscillate wildly between 0 and 1. An example of Schmitt trigger operation is shown in Figure 2.15 on next page.

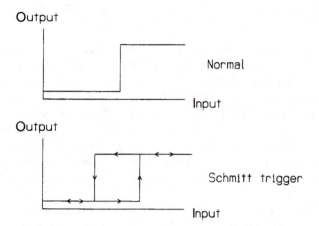

Figure 2.14 Schmitt Trigger

2.16 PROGRAMMABLE LOGIC DEVICES

Small logic circuits can be built with discrete or medium-scale integration components: for example, the 7400 family of logic components. These components implement various combinations of standard logic elements (ANDs, ORs, NORs, etc.), as well as some functional block elements such as counters. Large logic circuits, however, are not practical using these components because they would be much too large and use far too much

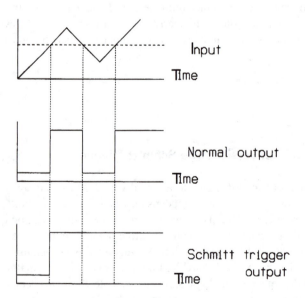

Figure 2.15 Schmitt Trigger Operation

power. Large-scale integration is the standard solution for components that will be produced in large numbers. Hundreds of thousands of gates can be integrated onto a single chip and manufactured at very low unit cost. Unfortunately, the development cost for integrated circuits is very high, so they are not practical for one or few-of-a-kind designs.

Simply building an integrated circuit with large numbers of AND, OR, NAND, . . ., gates is not a practical solution either. The problem there is not that the circuit could not be built; the problem is that each of the thousands of gates would need to have its input and outputs accessible from off the chip. This would lead to an impossibly large number of pins on the chip. The solution is to build chips that can implement arbitrary logic functions, rather than just copy circuits made from logic blocks. Called *programmable logic devices* (PLDs), these come in a number of configurations. The two families that are most important are the read-only memories (ROMs) and the programmable arrays (PALs and PLAs). In either case, an important property is how they are *programmed*, that is, what the physical process is that puts a specific logic function into a device. For small lot projects, such as prototyping, field-programmable devices are most desirable. In fact, for debugging, it is useful to have devices that are both field-programmable and erasable. These tend to be the most expensive, but the most flexible for laboratory work. At the other end of the spectrum are factory-programmed devices. These carry a significant setup cost but have the least manufacturing cost.

2.16.1 ROM as a Logic Device

Although ROMs are best known as computer memory components, they can also be used as logic components. A ROM is organized with a set of inputs specifying an *address*. For each address, there is a unique set of outputs, the *contents* of that address. The specific number of address inputs and the number of bits in the output vary from one unit to another. For a ROM with n inputs and m outputs, this can be viewed as m truth tables. That is, there are m output variables depending on the same input variables. Each output variable has its own truth table.

Another way to look at a ROM is as a sum-of-products representation of a function with no minimization; that is, all terms are included. This can be shown diagrammatically (see Figure 2.16 on next page). This is not a circuit diagram; it is a way of coding functional connections. The inputs show the variables and their complements (the complements are normally generated internally). Dots or small circles show permanent connections. In this context a *connection* is an input to to a multi-input gate. For a ROM the dots make the connections for all possible product terms (i.e., all the addresses are accessible). That means that the inputs are always connected to all possible sum-of-products terms. This is called a PLD with a fixed AND array.

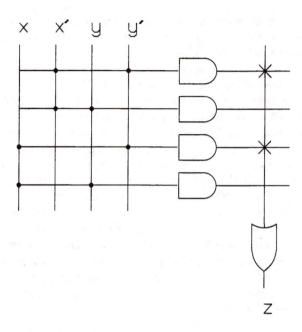

$$z = xy' + x'y'$$

Figure 2.16 PLD Format for a ROM, with an Example of an OR-Array Program

The Xs show variable connections, that is, programmable connections. This is where the device is programmed. In the case of the ROM, the OR-array is programmable. That is, while all possible sum-of-product terms are generated, only the ones specified in the programmable layer are actually used. For a two-input function as shown in the figure, there are four possible sum-of-product terms: xy, x´y, xy´, and x´y´. Any combination can be used. The figure shows the case for which the xy´ and x´y´ terms are used, so the ROM as shown implements the function,

$$z = x * y' + x' * y'$$

2.16.2 PAL, Programmable AND, Fixed OR; PLA, Both Programmable

Using a PLD with programmable AND and fixed OR arrays is more efficient because it allows for minimized sum-of-product forms. This arrangement was pioneered with PAL (Programmable Array Logic, a tradename of Monolithic Memories, since purchased by AMD) and is a very popular configuration. In this configuration the AND terms are connected via programmable elements, and the ORs are fixed to the outputs. Figure 2.17 shows a four-input, two-output PAL implementation of the function

$$z1 = x2 * x4 + x1 * x3$$
$$z2 = x2' * x3 * x4 + x1' * x2 * x3 * x4' + x1 * x2 * x3' + x1 * x4$$

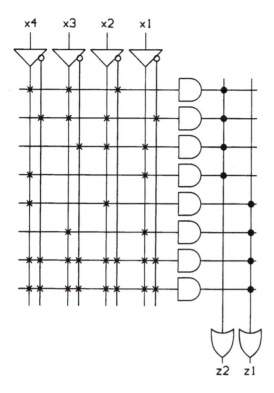

Figure 2.17 PAL Implementation of an Example Function

Note that the bottom two rows have all xs; this is the unprogrammed state and has no effect on the result.

The form of PLD that has both arrays programmable is called a PLA, programmable logic array. It adds more logic efficiency but at the expense of more internal complication and more pins. With the OR-array programmable, not only can minimized functions be used, but the AND rows can be assigned to any of the outputs. Figure 2.18 shows the same example as above implemented with a PLA. Note that the bottom two rows could be assigned to another output, such as z3 in the figure, and are ready to be programmed for an additional function. In general, the PLA architecture allows for more functions per chip.

2.16.3 Erasable PLDs

Field programmability and erasability have vastly widened the market for PLDs and generated a variety of devices with array logic at the core but surrounded with a variety of other elements. They now have greater functionality, so that complete logic systems

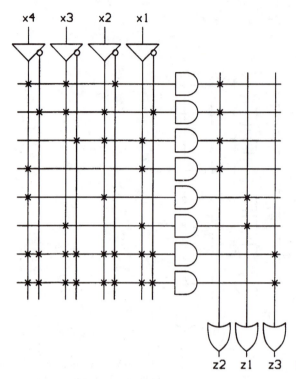

z2   z1   z3   **Figure 2.18 PLA Implementation**

(including sequential logic) can be built on a single chip. A number of standard software packages are available for programming these devices. The examples that follow use the software from Intel Corp. The general procedure is:

- Name inputs and outputs.
- Specify part to be used.
- Specify input and output connection types.
- Specify Boolean equations.

The program will simplify the expression (first putting it in sum-of-products form, if necessary), then produce programming information for the PLD. A second program can download that information to a PLD programmer for actual implementation.

As an example, the following set of functions is set up for programming an erasable PLD (EPLD). Both the input to the program and the output are shown.

$$z1 = x2 * x4 + x1 * x3$$
$$z2 = x2' * x3 * x4 + x1' * x2 * x3 * x4' + x1 * x2 * x3' + x1 * x4$$

Also, to show reduction ability, add the equation

$$L = (x5 + x6) * (x5' + x6) * (x5' + x6')$$

Note that this equation is not in sum-of-products form, so the program will convert it to that form and simplify it.

2.16.4 Logic Optimizing Compiler Input

The initial part of the input has identifying information and identification of the particular EPLD part to be programmed, as well as the kind of package it is in (so the program can correctly identify the input and output pins). In this case the part is a 5AC312, an industry standard small PLD.   The NETWORK section establishes the connections between inputs and outputs and the variables internal to the EPLD.

```
YOUR NAME
YOUR COMPANY
DATA
U1
001
5AC312
PACKAGE:  D5AC312-25
OPTIONS: TURBO=ON
PART: 5AC312

INPUTS:  x1,x2,x3,x4,x5,x6

OUTPUTS: z1,z2,L

NETWORK:
x1 = INP(x1)
x2 = INP(x2)
x3 = INP(x3)
x4 = INP(x4)
x5 = INP(x5)
x6 = INP(x6)
z1 = CONF(z1,vcc)
z2 = CONF(z2,vcc)
L = CONF(L,vcc)

EQUATIONS:

z1 = x2*x4+x1*x3;
z2= x2'*x3*x4+x1'*x2*x3*x4'+x1*x2*x3'+x1*x4;
L = (x5+x6)*(x5'+x6)*(x5'+x6');

END$
```

LOC Output: optimized equations

EQUATIONS:
*L = x5′ * x6;*

*z2 = x2 * x3 * x4′ * x1′*
  *+ x2 * x3′ * x1*
*+ x2′ * x3 * x4*
*+ x4 * x1;*

*z1 = x1 * x3*
*+ x2 * x4;*

END$

% FINAL LOC PIN ASSIGNMENTS
INPUTS:
 x1@3, x2@4, x3@5, x4@6, x5@7, x6@8

OUTPUTS:
 z1@14, z2@20, L@11

Note: the numbers following the @ are pin assignments

Chip utilization report

INTEL Logic Optimizing Compiler Utilization Report
IPLS II FIT Version 2.2 Level 4.1i 03/26/90

***** Design implemented successfully

YOUR NAME
YOUR COMPANY
DATA
U1
001
5AC312
PACKAGE: D5AC312-25
OPTIONS: TURBO=ON

5AC312
  - - - - -

Gnd -| 1 24|- Vcc
Gnd -| 2 23|- Gnd
x1 -| 3 22|- Gnd
x2 -| 4 21|- Gnd
x3 -| 5 20|- z2
x4 -| 6 19|- Gnd
x5 -| 7 18|- Gnd
x6 -| 8 17|- Gnd
Gnd -| 9 16|- Gnd

```
Gnd -|10    15|- Gnd
L -|11    14|- z1
GND -|12    13|- Gnd
    - - - - -
```

OUTPUTS

Name Pin Resource MCell PTerms | Sync Clock

```
z1   14     CONF      2    2/16 |    -
z2   20     CONF      7    4/16 |    -
L    11     CONF      1    1/12 |    -
```

INPUTS

Name Pin Resource MCell PTerms | Sync Clock

```
x1    3      INP       -     -  |    -
x2    4      INP       -     -  |    -
x3    5      INP       -     -  |    -
x4    6      INP       -     -  |    -
x5    7      INP       -     -  |    -
x6    8      INP       -     -  |    -
```

UNUSED RESOURCES

Name Pin Resource MCell PTerms

```
1    INPUT      -    -
2    MCELL     11   16
9    INPUT      -    -
10   INPUT      -    -
13   INPUT      -    -
15   MCELL      3   16
16   MCELL      4   16
17   MCELL      5   16
18   MCELL      6   12
19   MCELL      8   16
21   MCELL     10   16
22   MCELL      9   16
23   MCELL     12   12
```

PART UTILIZATION

```
3/12 MacroCells (25%), 15% of used Pterms Filled
6/10 Input Pins (60%)
PTerms Used 3%
```

Macrocell Interconnection Cross Reference

FEEDBACKS: *M M M*
0 0 0
1 2 7
L .... CONF @M1 -> . . . @11
z1 .... CONF @M2 -> . . . @14
z2 .... CONF @M7 -> . . . @20

INPUTS:

*x1 .... INP @3 -> . * **
*x2 .... INP @4 -> . * **
*x3 .... INP @5 -> . * **
*x4 .... INP @6 -> . * **
*x5 .... INP @7 -> * . .*
*x6 .... INP @8 -> * . .*
                  L z z
                  1 2

. = not connected   x = no connection possible
* = signal feeds cell   ? = error, unable to fit

The *macro cells* referred to above are internal partitions that are used to make efficient use of the chip for lots of small equations. Macro cells allow for some interconnect,
giving some but not all of the functionality of PLAs but with less complexity. There are
many other ways to connect the outputs; these will be exploited in using EPLDs for
sequential logic.

2.17 PROBLEMS AND DISCUSSION TOPICS

1. The switches shown in Figure 2.19 represent a logic circuit with four inputs, one for each
    switch. When the appropriate combination of switches is closed, the bulb will light. Consider
    a closed switch to be described as a "1" value of the associated variable, and the output of the
    circuit to be "1" when the bulb is lit.

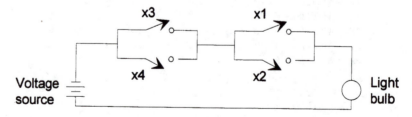

Figure 2.19 Switch-based logic circuit

(a) Derive the truth table for this system.
(b) Use the truth table to derive an equivalent gate circuit (based on AND, OR, and NOT
    operators).

  (c) Use the truth table and algebraic reduction to derive sum-of-products and product-of-sums forms for the Boolean solution to this problem. Draw switch-based circuits for each and compare to the original circuit.

2. Build a truth table to prove the distributivity property of Boolean algebra. For example, show through the use of a truth table that $x + (y * z) = (x + y) * (x + y)$.

3. A three-bit absolute encoder is shown in Figure 2.20. The white regions represent 0 and the dark regions represent 1 (this convention is arbitrary). As the wheel rotates clockwise, a natural binary sequence is generated, with the outer ring generating the least signficant bit and the inner ring the most significant bit. The sequence will read 000, 001, 010, and so on.

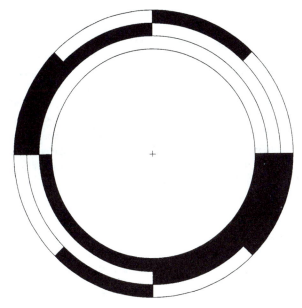

Figure 2.20 Absolute Encoder

  (a) Verify that the rest of the binary sequence is generated correctly.
  (b) Identify any transitions in this sequence that produce hazards (more than one bit changing at a time).
  (c) Design an equivalent Gray code encoder wheel that does not have any hazards.
  (d) Design a Boolean circuit that will convert the Gray code to natural binary.

4. The two-output logic circuit shown in Figure 2.21 is built from AND, OR, and NOT gates.
  (a) Derive the algebraic expressions for $z1$ and $z2$.
  (b) Generate the truth table from the algebraic expression.
  (c) Work directly from the circuit diagram to write a computer program (in any convenient language) to verify the truth table generated in part (b). It is easiest to define names for the intermediate results, paying careful attention to computing order.
  (d) Derive minimized circuits for the two outputs from the truth tables. Compare the cost of these minimized circuits to the original (the number of gates is a reasonable rough estimate of cost). How much expense is added to the circuits to make them hazard free?
  (e) Convert the original circuit and the minimized circuits so that they are constructed only with NAND gates, then only with NOR gates.

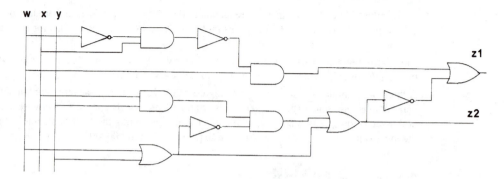

Figure 2.21 Two-Output Logic Circuit

5. Use a photodiode as a logic input element with a Schmitt trigger to clean up the output. Using a fast digital oscilloscope, compare the output of the photodiode to the output of the Schmitt trigger. Use the photodiode circuits to build a typical machine safety switch. It will produce an ON signal only if two fingers on each hand are used to block light to four photo diodes.

6. Construct the circuit of Problem 4 using a ROM, a PLD, and discrete gates. Experiment with several input devices for w, x, and y: photodiodes, mechanical switches, computer outputs, and so on. In each case, examine the quality of the input signal in terms of clean transitions, presence or absence of ringing, consistent levels, and so on.

3

Synchronous Sequential Logic

When a system requires that the output depend on the history of the inputs as well as the present value, a *sequential circuit* is needed. Most of the interesting problems in digital logic are sequential rather than purely Boolean. The history dependence in a logic circuit is achieved by using a Boolean circuit with signals from the output fed back to the input (Figure 3.1). The feedback signals shown in the figure are known as *state variables*. They contain the information about system history that is needed to produce the correct sequences. At any given moment, the combination of state information and input information is used to determine the state for the next moment. That change in state is called a *transition*. When the state variables are Boolean, there are only a finite number of states that the system could possibly exhibit. The name *finite-state machine* is applied to such systems.

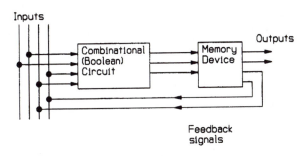

Figure 3.1 Sequential Logic Circuit

3.1 FLIP-FLOPS

The simplest memory element is the time delay caused by processing speed limitations of the logic devices. Explicit memory elements make circuit design easier and are more commonly used than delays. The set–reset (SR) flip-flop is very common; it "remembers" the most recent input on its S or R input lines. The SR flip-flop is itself a sequential circuit that uses delays for its memory elements, shown in Figure 3.2. The lines labeled S and R are the inputs; the lines labeled QP and Q are the outputs. QP is the complement of Q.

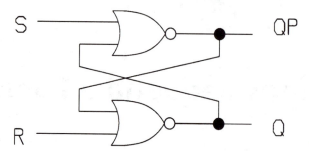

S

QP

R

Q

Figure 3.2 Set–Reset Flip-Flop

An intuitive walk-through of its behavior can give an idea of how it is used. Assume that, to start, S, R, and Q are 0 and QP is 1. In this state the input to the upper NOR gate is (S,Q) = (0,0), so its output, QP, is 1. QP is thus stable. The input to the lower NOR gate is (R,QP) = (0,1), so its output, Q, is 0. Q is thus also stable. As long as the inputs are not changed, the outputs will maintain their initial values.

If S is changed to 1, the input to the upper NOR becomes (1,0) and its output, QP, changes to 0. The input to the lower NOR becomes (0,0), so its output, Q, changes to 1. This then changes the input to the upper NOR to (1,1), but that does not change its output, so QP stays at 0. The circuit has thus reached a stable state, with Q = 1 and QP = 0. If S is then changed back to 0, the input to the upper NOR changes to (0,1). This does not change its output, however, so the system reaches a new stable state with (S,R) = (0,0) and (Q,QP) = (1,0). In the new state, the inputs are again (0,0), but the output has changed to Q = 1. The fact that S was momentarily changed to 1 has thus been "remembered."

The same procedure can be used to show that if R is now changed to 1, the output will change to Q = 0, QP = 1, and if R is then changed back to 0, the output will not change. For this implementation, if S and R are both 1, Q and QP will both be 0, which violates the design goal that QP is the complement of Q. Thus (S,R) = (1,1) is considered an invalid input. Other implementation of the SR flip-flop can have other behaviors for this invalid condition, including oscillation.

A simple computer program (in this case, in C) can simulate its behavior:

```
#include <stdio.h>

#define NSR 14
static int ss[] = {0,1,1,1,0,0,0,0,0,0,0,1,1,1};
static int rr[] = {0,0,0,0,0,0,1,1,1,0,0,1,1,1};
static int n = 13;
```

```
main(ac,av)
int ac;
char *av[];


{
int s,r,q = 0,qp = 1,nxt_q,nxt_qp,i;

printf("s r   q qp   nq nqp\n");

for(i = 0; i < n; i++)

{
if(i >= NSR)
{
s = ss[NSR - 1];
r = rr[NSR - 1];
}
else
{
s = ss[i];
r = rr[i];
}
nxt_q = !(qp || r);

nxt_qp = !(q || s);
printf("%d %d   %d %d    %d   %d\n",s,r,q,qp,nxt_q,nxt_qp);
q = nxt_q;
qp = nxt_qp;
}
}
```

The arrays ss[] and rr[] are values of the S and R inputs at each time step. The variables nxt_q and nxt_qp are the outputs of the Boolean portion of the circuit. They are the next values that the outputs will have. Note that it is important to distinguish the next values from the current values (q and qp) because the current values are used in the Boolean circuit. They are the feedbacks that make this a sequential circuit. The results of running this program are as follows:

```
s r   q qp   nq nqp
0 0   0 1    0  1
1 0   0 1    0  0
1 0   0 0    1  0
1 0   1 0    1  0
0 0   1 0    1  0
0 0   1 0    1  0
0 1   1 0    0  0
0 1   0 0    0  1
0 1   0 1    0  1
0 0   0 1    0  1
```

```
00 01 0 1
11 01 0 0
11 00 0 0
```

By reviewing the output characteristics, the following operating rules for an SR flip-flop can be verified:

- The two outputs (q and qp) are complements.
- When S = 1, Q -> 1.
- When SR = 00, Q does not change.
- When R = 1, Q -> 0.

This new element can now be used to define a more sophisticated door lock, one that requires the right sequence of button pushes. This provides many more combinations for the same number of buttons. Boolean logic could not have been used to design this because it is an algebra and has no way to represent sequences. The properties of static and dynamic digital systems and the parallel properties of systems in general are summarized in Table 3.1.

TABLE 3.1   PROPERTIES OF STATIC AND DYNAMIC SYSTEMS

	Static systems	Dynamic systems
Systems in general	Algebraic equations	Differential equations; real state variables
Boolean logic systems	Combinational (Boolean) logic	Sequential logic; Boolean state variables

The SR flip-flop, with its feedback, can represent sequences. For the new design, the unlock function will require that the user press and release button 1, then press and release button 2. Figure 3.3 shows a correct and an incorrect sequence of pushes. A brute-force design for this system can be established by identifying the unique conditions associated with each step. Each condition is called a *state*. As noted from Figure 3.3, there are four unique conditions or states, not counting the initial state. These are when button 1 is on but button 2 is off, then both are off, then button 2 is on while 1 is off, and finally, both are off again. Each of these states can be associated with an SR flip-flop; being in a state is associated with the corresponding flip-flop being on. The general design for four flip-flops is shown in Figure 3.4. The design problem, then, is to figure out what Boolean circuits go in the excitation blocks for each flip-flop.

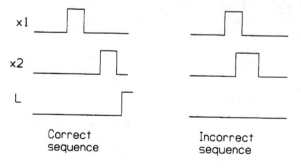

Correct
sequence

Incorrect
sequence

Figure 3.3 Door Lock Button Sequence

x1 x2

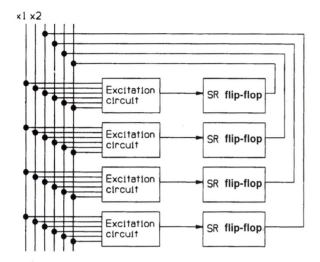

Figure 3.4 General Layout for Sequential Lock Design

The idea is to get each flip-flop to turn on in sequence as the proper button sequence is pushed. The lock opens when the final flip-flop turns on. For the first flip-flop, for example, x1 must be 1 while x2 and Q1 are 0 for it to turn on (i.e., s1 = 1), indicating the start of the sequence. Following this reasoning, the excitation equations are

$$s1 = x1 * x2' * q1'$$
$$s2 = x1' * x2' * q1$$
$$s3 = x1' * x2 * q2$$
$$s4 = x1' * x2' * q3$$

The output equation is

$$L = q4$$

Note that R (reset) inputs have not been used at all; they could be used as master reset. As before, this can be simulated with a computer program.

```
#include <stdio.h>

#define NX 11
static int xx1[] = {0,1,1,1,1,1,1,0,0,0,0};
static int xx2[] = {0,0,0,0,0,0,1,1,1,0,0};

int sr(int s,int r,int q)/* Return next value for sr flipflop output
_____*/
{
int next;

if(s && r)
{
printf("Illegal SR Flipflop state\n");
exit(1);
}
```

```
if(s)next = 1;
else if(r)next = 0;
else next = q;
return(next);
}

main()
{
int x1,x2,L;
int s1,s2,s3,s4;
int r1 = 0,r2 = 0,r3 = 0,r4 = 0;
int q1 = 0,q2 = 0,q3 = 0,q4 = 0;
int nq1,nq2,nq3,nq4;
int i,n = 13;

printf("x    s       q      next-q   L\n");

for(i = 0; i < n; i++)
{
if(i >= NX)
{
x1 = xx1[NX - 1];
x2 = xx2[NX - 1];
}
else
{
x1 = xx1[i];
x2 = xx2[i];
}
s1 = x1 && !x2 && !q1;
s2 = !x1 && !x2 && q1;
s3 = !x1 && x2 && q2;
s4 = !x1 && !x2 && q3;
L = q4;
nq1 = sr(s1,r1,q1);
nq2 = sr(s2,r2,q2);
nq3 = sr(s3,r3,q3);
nq4 = sr(s4,r4,q4);
printf("%d %d  %d %d %d %d  %d %d %d %d  %d %d %d %d   %d\n",
x1,x2,s1,s2,s3,s4,q1,q2,q3,q4,nq1,nq2,nq3,nq4,L);
q1 = nq1;
q2 = nq2;
q3 = nq3;
q4 = nq4;
}
}
```

The results of this simulation are shown on the following page for two different sequences, the first correct, the second incorrect.

x	s	q	next-q	L
00	0000	0000	0000	0
10	1000	0000	1000	0
10	0000	1000	1000	0
00	0100	1000	1100	0
00	0100	1100	1100	0
00	0100	1100	1100	0
01	0010	1100	1110	0
01	0010	1110	1110	0
01	0010	1110	1110	0
00	0101	1110	1111	0
00	0101	1111	1111	1
00	0101	1111	1111	1
00	0101	1111	1111	1

x	s	q	next-q	L
00	0000	0000	0000	0
10	1000	0000	1000	0
10	0000	1000	1000	0
10	0000	1000	1000	0
10	0000	1000	1000	0
10	0000	1000	1000	0
11	0000	1000	1000	0
01	0000	1000	1000	0
01	0000	1000	1000	0
00	0100	1000	1100	0
00	0100	1100	1100	0
00	0100	1100	1100	0
00	0100	1100	1100	0

But will it work? Probably. The first doubts are at the design level. Have all possible sequences been considered? Is there an incorrect sequence that will open the lock anyway? There could be. The design of the state logic considered only the states of the adjoining flip-flops. To make sure that the system is really in its initial state, for example, all four flip-flops should have zero outputs. Then there are questions of timing. The computer program imposes the timing of the sequential statement execution. The actual circuit components, however, will change state as fast as they can, in parallel. This design is straightforward and uses cascaded flip-flops, so timing will probably not be a problem. The bottom line, however, is that there is no systematic way to check the design, to check for hazards in the Boolean circuits, to minimize the circuits, and so on. This ad hoc design method has enough limitations so that it is clearly not useful for any but the simplest sequential logic designs.

3.2 SYNCHRONOUS SYSTEMS

A major source of errors in sequential circuits are the time delays, caused by the finite time required for gates to change state. These are present in the excitation circuits and therefore in the time it takes feedback signals to arrive. In synchronous circuits, a clock is used to make sure that all signal propagation delays are covered, so that no changes can take place while signals are propagating. This provides the same noise margin in the temporal domain that binary quantization provides in the signal domain. Timing constraints become strictly local to each memory element. That is, each element's timing characteristics must meet the specifications of the master clock and are not in any way dependent on the characteristics of other circuit elements. The clock restricts all input circuitry to be stable before memory elements are allowed to change. Any asynchronous system inputs should be passed through synchronizers before being presented as inputs to the synchronous part of the circuit. This is so that the same signal will not have different values in different parts of the circuit.

Design of a synchronous system starts with a state transition diagram. This diagram is the connection between the engineering problem description and the logic circuit. The state diagram is usually constructed by expanding the design statement and adding the details necessary to make a rigorous logic specification. The state diagram formulation also has application beyond sequential logic and is often used for software design and description of other problems in which sequence is a dominant factor.

3.3 STATE TRANSITION LOGIC

Figure 3.5 shows the transition logic diagram for the lock. The circles represent states. The text at the top of the circle is the state name; the text below that is the output value associated with that state. The arrows are transitions from one state to another. The numbers above the arrows give the input values that will enable the transition to take place. Until an input combination satisfying one of the transitions from a state occurs, the system stays in its present state. This is sometimes shown with self-transition arrows, but these have been omitted from this diagram to simplify it.

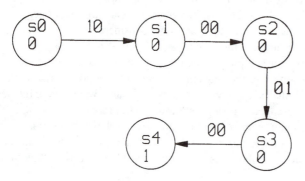

Figure 3.5 Transition Logic Diagram for Door Lock

3.4 NEXT-STATE TABLE

A *next-state table* is constructed directly from the state transition diagram. This is a purely formal step in the sense that it represents exactly the same information in a different form (Table 3.2). To read the table, find the row corresponding to the current state. The column headings are for all possible combinations of input variables. To find the next state, locate the column corresponding to the current input values, and read the next state from the row corresponding to the current state.

TABLE 3.2 NEXT-STATE TABLE

	\multicolumn{4}{c}{Inputs, x1 x2}			
	00	01	11	10		
	S0	S0	S0	S0	S1	
Current state	S1	S2	S1	S1	S1	
	S2	S2	S3	S2	S2	
	S3	S4	S3	S3	S3	
	S4	S4	S4	S4	S4	

3.5 STATE ASSIGNMENT

The number of states is at least $\log 2n$, where n is the number of states. In this case there are at least three state variables, although some minimization might be possible later. State assignment is the process of associating a set of logic variable values with each of the named states in the next-state table. Synchronous circuits are not very sensitive to state assignment, so there is no fixed rule for this. In this case an assignment using Gray codes can be made:

S0 000
S1 001
S2 011
S3 010
S4 110

3.6 MEMORY ELEMENTS FOR SYNCHRONOUS CIRCUITS: D FLIP-FLOPS

In synchronous circuits the memory elements must be clocked so that excitation circuitry can settle before the memory elements change. To avoid oscillation, devices are normally edge triggered: that is, outputs change on arrival of a clock transition. The easiest memory element to design with is the D flip-flop. It passes input to output on clock transition. At all other times, the output is held constant. Note that SR flip-flops cannot be used in synchronous systems because they have no provision for a clock input. The next-state description for a D flip-flop is simply

$$Q^+ = Q$$

The D flip-flop is shown on next page in schematic form in Figure 3.6. In electrical terms, the next-state description means that the output, Q´, becomes equal to the input, D, at the sampling moment determined by the clock (or control) input, C. The complementary output, Q´, is always equal to the complement of Q. In most cases, that moment is defined as the time at which C changes from 0 to 1. This is called an *edge-sensitive flip-flop*. The shape of the clock signal does not matter in this case because it is the change in value of

C that causes the flip-flop to be activated. The input signal, D, must be stable (constant) during the short period around the transition time; otherwise, the output will be unpredictable.

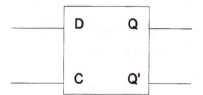

Figure 3.6 D Flip-Flop

D flip-flops can also be designed for which the output changes while the clock is 1. To avoid oscillatory behavior for this type of design, the clock signal is normally made up of very short pulses, pulses so short that the sampling moment is comparable to the delay times of the gates making up the flip-flop. Again, for consistent results, the input, D, must be stable for the entire time that the clock is 1. When edge-sensitive (also called *edge-triggered*) flip-flops are used, the most common clock signal is a square wave rather than the sequence of very narrow pulses that must be used with a clocked flip-flop.

To design with a D flip-flop, the next-state diagram is combined with the state assignment to produce excitation equations. First, modify the next-state table so that the actual state-variable values are used instead of the state names (Table 3.3). Then excitation equations or maps can be written directly from this table by looking at the Boolean relations that are needed to get the next state from the present state. The excitation map to get q1+ is shown in Figure 3.7. By making all of the don't cares 1, this gives the excitation equation for q1+ as

TABLE 3.3 NEXT-STATE TABLE, Q1+,Q2+,Q3+

		Inputs, x1 x2			
		00	01	11	10
Current	000	000	000	000	00
state,	001	011	001	001	001
q1,q2,q3	011	011	010	011	011
	010	110	010	010	010
	110	110	110	110	110

$$q1^+ = x1' * x2' * q2 * q3' + q1$$

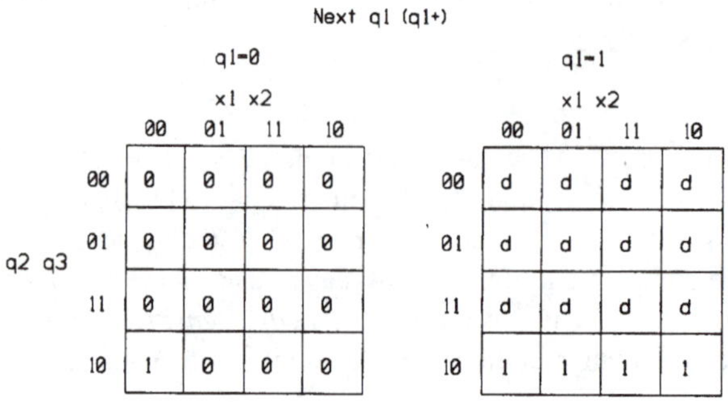

Next q1 (q1+)

q1=0						q1=1			
	x1 x2						x1 x2		
	00	01	11	10		00	01	11	10
00	0	0	0	0	00	d	d	d	d
01	0	0	0	0	01	d	d	d	d
11	0	0	0	0	11	d	d	d	d
10	1	0	0	0	10	1	1	1	1

q2 q3

Figure 3.7 Door Lock Excitation for q1$^+$

Maps for q2+ and q3+ are shown in Figures 3.8 and 3.9. The excitation equations are thus

$$q2^+ = x1' * x2' * q3 + q2 + q1$$
$$q3^+ = x1 * x2' * q2' + x1 * q3 + q2' * q3 + x1' * x2' * q3$$

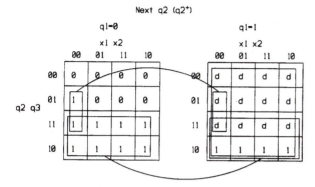

Figure 3.8 Excitation Map for q2+

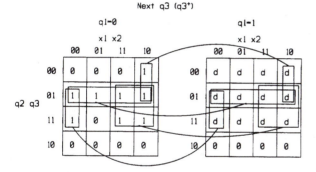

Figure 3.9 Excitation Map for q3+

The output is on only when the system is in state 4 (s4). S4 = 110, so the output is

$$L = q1 * q2 * q3'$$

Again, this can be simulated.

```
/* Sequential door lock using D-flipflops */

#include <stdio.h>

#define NX 11
static int xx1[] = {0,1,1,1,0,0,0,0,0,0,0};
static int xx2[] = {0,0,0,0,0,0,1,1,1,0,0};

...
/* Excitation equations */

nq1 = (!x1 && !x2 && q2 && !q3) || q1;
nq2 = (!x1 && !x2 && q3) || q2 || q1;
nq3 = (x1 && !x2 && !q2) || (x1 && q3) || (!q2 && q3) ||
(!x1 && !x2 && q3);
```

```
L = q1 && q2 && !q3;
printf("%d %d   %d %d %d  %d %d %d   %d\n",
x1,x2,q1,q2,q3,nq1,nq2,nq3,L);
q1 = nq1;
q2 = nq2;
q3 = nq3;
}
}
```

The results for a "correct" sequence are:

```
x   q     next-q   L
00  000 000  0
10  000 001  0
10  001 001  0
10  001 001  0
00  001 011  0
00  011 011  0
01  011 010  0
01  010 010  0
01  010 010  0
00  010 110  0
00  110 110  1
00  110 110  1
00  110 110  1
```

Will this version work? Almost certainly. The design is consistent and can be checked against the original specifications. Because of the clock there are no timing problems — the actual circuit behaves very much like the computer program in the sense that all inputs change, then all outputs, and so on. The only timing constraint is that the clock should not be too fast. The clock speed must be slow enough so that all the logic elements have time to settle before the clock changes again. For standard logic devices, the maximum speed is about 50 to 100 MHz. More than fast enough for this problem. This design is not complete, however. For example, there is no way to reset, either after successful or unsuccessful usage. There should probably also be an output to the user, perhaps a light, to indicate the system state.

3.7 OTHER SYNCHRONOUS MEMORY ELEMENTS

Several other types of flip-flops can be used in synchronous circuits. The JK flip-flop is equivalent to an SR flip-flop with the addition of a clock input. In addition, when J and K are both on, the output will toggle. A T flip-flop has only one input; when that input is on, the output toggles. A master–slave, edge-triggered flip-flop is a two-stage device. The second stage operates off an inverted clock. The input propagates through the master–slave flip-flop in a two-step process, thereby preventing the input from directly affecting the output. It is more complicated internally but affords an extra degree of synchronous protection.

The terms *latch, flip-flop, memory*, and *register* are used almost interchangeably. There are some distinctions, which, even when the language is ambiguous, should be borne in mind in terms of functions. All of these are clocked elements. A latch is an element for which the output changes with the input as long as the clock is on. When the clock is off, the latch holds its last value. The term *transparent latch* is sometimes used to distinguish it from the ambiguous use of the term *latch*. A flip-flop's output changes only when the clock changes (i.e., on a clock transition). The terms *memory* and *register* are used for the same function. A memory holds the value that was presented to it at the last clock transition. A D flip-flop is a one-bit register. Multibit registers, which are the most common usage, can be made by combining D flip-flops.

To use a flip-flop other than a D-type, the next-state table is used. The excitation must be designed based on the characteristics of the particular flip-flop. For example, for a T flip-flop, the excitation input will be 1 if the next state is different than the present state; otherwise it will be 0. The T flip-flop is often considered the most economical to use. The D will be used here more frequently because of the ease of designing with it, but as an example, the excitation development for the sequential door lock is done below using a T flip-flop.

As before, the procedure starts with the next-state table, which is repeated in Table 3.4 for the door lock. The excitation map for q2 is shown in Figure 10. This map has only a single 1 entry; thus the excitation function is extremely simple. This is true for the other excitation maps as well. In this case, the excitation circuitry is certainly simpler for a T flip-flop than for a D flip-flop.

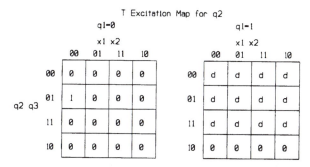

Figure 3.10 T-Flip-Flop Excitation Map for Door Lock

TABLE 3.4 NEXT-STATE TABLE, Q1+,Q2+,Q3+

	Inputs, x1 x2				
	00	01	11	10	
Current	000	000	000	000	001
state,	001	011	001	001	001
q1,q2,q3	011	011	010	011	011
	010	110	010	010	010
	110	110	110	110	110

3.8 USING EPLDS FOR STATE MACHINES

As mentioned in Chapter 2, PLDs have been enhanced to allow complete logic system implementation in a single chip. The example below shows implementation of the door lock design using two different Intel EPLDs, one that includes D flip-flops (85C220 and 85C224) and one that includes T flip-flops (5AC312). The implementation is done using the Intel design software, which starts with the input of the transition logic and goes all the way through to loading the program into the chip (a number of other software design tools will also perform substantially the same functions as the Intel software used here). The state-transition information is put in IF-THEN form, from which all the Boolean excitation information is developed. The listings below show the program inputs and outputs for each stage. This procedure is similar to that used by many other PLD suppliers and software packages, some of which are more elaborate, some less.

State machine input for door lock

```
YOUR NAME
YOUR COMPANY
DATE
1
A
85C224
LOCK2:  Door Lock


OPTIONS:  TURBO = OFF
PART:  85C224
INPUTS:
CLK
x1
x2
OUTPUTS:
L
Q1% This state variable will appear as an output
—others will not %
NETWORK:
L = CONF(L,VCC)


EQUATIONS:


L = Q1 * Q2 * Q3';


MACHINE:  LOCK2
CLOCK:  CLK
STATES:   [ Q1   Q2   Q3]
S0       [ 0    0    0]
S1       [ 0    0    1]
S2       [ 0    1    1]
S3       [ 0    1    0]
S4       [ 1    1    0]


S0:
IF x1 * x2' THEN S1
S1:
```

```
IF x1' * x2' THEN S2
S2:
IF x1' * x2 THEN S3
S3:
IF x1' * x2' THEN S4
S4:
IF x1 * x2 THEN S0


% The state machine software will not
allow a state with no transitions.  This will
go back to start when both buttons are pressed %

END$
```

Logic design for lock

```
...
NETWORK:
CLK = INP(CLK)
x1 = INP(x1)
x2 = INP(x2)
L = CONF(L,VCC)

%
I/O's for State Machine "LOCK2"
%
Q1, Q1 = RORF(Q1.d, CLK, GND, GND, VCC)
Q2 = NORF(Q2.d, CLK, GND, GND)
Q3 = NORF(Q3.d, CLK, GND, GND)

EQUATIONS:
L = Q1 * Q2 * Q3';
%
Boolean Equations for State Machine "LOCK2"
Current State Equations for "LOCK2"
%
S0 = Q1'*Q2'*Q3';
S1 = Q1'*Q2'*Q3;
S2 = Q1'*Q2*Q3;
S3 = Q1'*Q2*Q3';
S4 = Q1*Q2*Q3';
%
SV Defining Equations for State Machine "LOCK2"
%
Q1.d = S4.n;
Q2.d' = S1.n
     + S0.n;
Q3.d = S1.n
    + S2.n;
%
```

Next-State Equations for State Machine "LOCK2"
%
S1.n = (S1 * (x1' * x2')')
*   + (S0 * (x1 * x2'));*
S0.n = (S4 * (x1 * x2))
*   + (S0 * (x1 * x2')');*
S2.n = (S2 * (x1' * x2)')
*   + (S1 * (x1' * x2'));*
S4.n = (S4 * (x1 * x2)')
*   + (S3 * (x1' * x2'));*

END$

Minimized design

...
EQUATIONS:
Q3.d = Q1' * Q2' * x1 * x2'
*   + Q1' * Q3 * x2'*
*   + Q1' * Q2' * Q3*
*   + Q1' * Q3 * x1;*

Q2.d' = Q1 * Q2 * Q3' * x1 * x2
*   + Q1' * Q2' * Q3'*
*   + Q1' * Q2' * x2*
*   + Q1' * Q2' * x1;*

Q1.d = Q2 * Q3' * x1' * x2'
*   + Q1 * Q2 * Q3' * x2'*
*   + Q1 * Q2 * Q3' * x1';*

L = Q1 * Q2 * Q3';

END$

The same problem can be solved with a device using T flip-flops (5AC312). The solution is shown below (the minimized equations).

EQUATIONS:
Q3.t = Q1' * Q2' * Q3' * x1 * x2'
*   + Q1' * Q2 * Q3 * x1' * x2;*

Q2.t = Q1' * Q2' * Q3 * x1' * x2'
*   + Q1 * Q2 * Q3' * x1 * x2;*

Q1.t = Q1' * Q2 * Q3' * x1' * x2'
*   + Q1 * Q2 * Q3' * x1 * x2;*

L = Q1 * Q2 * Q3';

Note that these excitation equations are much simpler than those for the D flip-flop.

As we did previously with a C program, these results can be simulated, but this time using a program that is part of the design package. The input to the simulation is a vector file giving the input values for each clock tick, and the output is a diagram of the states and outputs as a function of time.

;Vector file for LOCK2 simulation

```
000
100
010
110
000
100
000
101
001
101
000
100
000
000
```

The results of the simulation are shown below.

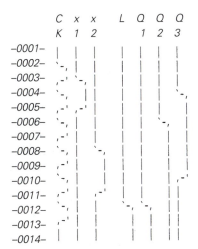

3.9 ASYNCHRONOUS INPUTS

There are many situations when an input signal to a logic circuit is produced by an element that is not synchronized by the clock. For example, if a switch closure caused by mechanical motion is an input, the switch is an *asynchronous* element. The input can change at any time in the clock cycle. Although asynchronous inputs are rare in very large synchronous systems such as computers, they are very common in mechanical control situations. The inputs from an incremental encoder—for example, a very common position-measuring element in precision control—are completely asynchronous with the logic circuitry and can have quite high frequency, up to 5 or 10 MHz, so generate large numbers of asynchronous events.

If, as is usual, such an input enters the system through a bus and is then distributed to all places in the circuit where it is used, inconsistent results could be obtained if the signal changes too closely to a clock transition. Depending on the delays in each of the paths by which it enters the circuit, it could appear to have different values in different circuit sections. This would produce erroneous results. The best solution to this is first to pass the signal through a clocked D flip-flop, then use the output of the flip-flop as the input variable for the circuit. This will solve most such timing problems since the output of the D flip-flop is synchronized.

3.10 MOORE–MEALY MACHINES

The synchronous circuits discussed thus far have had outputs that are functions only of the states, not the inputs. These are called Moore machines. Mealy machines have outputs that are functions of the states and inputs and, possibly, the clock. In transition diagrams, Moore outputs are associated with state bubbles while Mealy outputs are associated with the transitions. Systems that require pulse outputs are often designed as Mealy machines. Moore machines are easier to design and will be the only synchronous systems treated here.

3.11 STATE TABLE REDUCTION

It might be possible to reduce the number of rows in a state table. In some cases the number of state variables might decrease, and in most cases, the excitation circuitry will get simpler. When using random logic, this will result in a smaller system. Using PLDs, it might allow an otherwise-too-big design to fit into a PLD, allow the use of a cheaper PLD, and so on. In some cases, though, it will not make any difference at all. A manual minimization process is described here. Logic design software will often use this and other methods for minimization.

The reduction process uses the concept of equivalent rows. These are rows that, effectively, contain the same sequential information, so can be combined. The first rule for equivalence is that rows with different outputs cannot be combined. (This eliminates most reduction in Mealy machines because output depends on the inputs also, so can be different for each column.) The remaining state table rows are then examined to see if there is any hidden redundancy.

The first step is to construct an implication table. It has states on each axis; the box at the intersection of states is used to indicate the conditions for compatibility. The implication table for the door lock is shown in Figure 3.11. The first entry shows the initial information. Any boxes with Xs are known to be nonequivalent because their outputs conflict. For other boxes, the equivalency conditions are indicated. These are other pairs of states that must be equivalent if the test pair is to be equivalent. For example, the entry in the upper left position reads "0,2." This position is the equivalency test for state 0 (S0) to be equivalent to state 1 (S1). Looking in the first column of Table 3.5 state 0 has a 0 for its next state, while state 1 has a 2 for its next state. Thus for these two states to be equivalent, the states to which they are going, 0 and 2, must also be equivalent.

The next steps are a series of passes through the implication table. For each pass, the conditional pairs are checked. If the conditional pair has an X, the test fails and an X is placed in the test box. The procedure is completed when a full pass is made with no new Xs being placed. The remaining sections of the figure show these passes. In this case, no equivalent states were discovered so no reduction can be made. If any equivalent states were discovered, the rows corresponding to those states in the next-state table are combined, giving a new next-state table. The design then proceeds as before.

TABLE 3.5 NEXT-STATE TABLE:

		Inputs, x1x2			
		00	01	11	10
Current	S0	S0	S0	S0	S1
state	S1	S2	S1	S1	S1
	S2	S2	S3	S2	S2
	S3	S4	S3	S3	S3
	S4	S4	S4	S4	S4

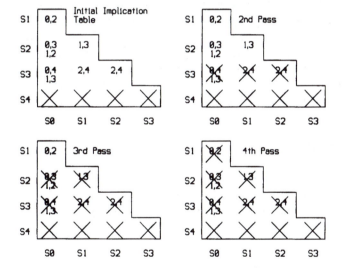

Figure 3.11 Implication Table for Door Lock

3.12 METASTABILITY*

Bistable devices have two stable equilibria, the 0 position and the 1 position. Between them, there is also at least one unstable equilibrium. This unstable equilibrium can cause problems in sequential logic systems. If an input happens to have the right energy content to put the system very near this unstable equilibrium, the time it will take to complete the transition, either to the opposite state or back to the original state, is unpredictable and can cause circuit timing errors.

*This discussion is based on Intel Application Note AP-336, *Metastability Characteristics of Intel EPLDs*, Thom Bowns, Intel Corporation, Santa Clara, California.

Timing characteristics of sequential circuits are based on two times: setup time (T_{su}) and output delay time (T_{co}). In synchronous systems, the clock period is chosen to be longer than the sum of these plus a safety margin. Metastability can occur when T_{su} is violated. This means that an input changes less than T_{su} before a clock edge. When that happens, the device could come close to its unstable equilibrium, and could stay there for a relatively long time. This would violate T_{co}, which means that the circuit is no longer synchronous. The result is then unpredictable.

The potential for metastable behavior in synchronous sequential systems exists whenever there is an asynchronous input. If a D flip-flop is used to synchronize an asynchronous input, the metastability problem focuses on that flip-flop. A "just-right" (or just-wrong, depending on your point-of-view) input timing can put the D flip-flop into a metastable state, in effect rendering its output an asynchronous rather than a synchronous signal. This could nullify the reason for which the D flip-flop was used in the first place. The probability of metastable behavior entering the synchronous part of the system can be reduced by using two D flip-flops in series, although there would be a cost penalty because of the extra components and a speed penalty because of the extra delay involved. There are no metastability problems in the interior of synchronous circuits, since all changes are guaranteed to occur with the clock. Probability calculations can be made on most logic components, so the extent of the danger can be predicted.

In asynchronous circuits (treated in Chapter 4), the rule is that inputs must change one at a time. Metastable behavior can cause this rule to be violated, also causing unpredictable behavior. Metastable behavior is inherent to logic components. There is no way to design systems that do not have potential metastability problems. At best, the probability of metastability can be minimized. From the view of the "perfect" nature of digital systems, metastability problems are the temporal equivalent of the probability that noise in a logic circuit will exceed the buffer-zone width. The probabilities can be made acceptably low, but neither is impossible.

3.13 PROBLEMS AND DISCUSSION TOPICS

1. Implement both the ad-hoc and the systematic solutions given for the door-lock problem at the beginning of the chapter. Test the performance and reliability of each.

2. Design a circuit that will change its output on every third pulse of an input signal (see Figure 3.12). Base the design on a synchronous circuit with an independent clock.

Figure 3.12 Input–Output Signals, Problem 2

Implement the circuit with individual logic components, with a ROM, and with a PLD. Examine the behavior of the circuit as the input pulse frequency increases toward the clock frequency.

3. Redesign the circuit of Problem 2 using the input signal as the clock. This method is often used to synchronize a command signal with a rotating shaft or other mechanical component.

4. Counting is a major logic activity in many applications. Design and implement a four-bit counter that counts the number of leading edges (0-to-1 transitions) of a pulse train. Use an independent clock.

5. Design a circuit that counts all edges of an input signal.

    Signals consisting of 90° out-of-phase square waves are called *quadrature* signals. They are used in a variety of applications, including stepping motor control and incremental encoders. The next two problems deal with signals of this sort.

6. Given two input signals, design a circuit that counts all edges of both signals, but only if the edges alternate from one input to the other. As soon as alternation stops (i.e., as soon as two consecutive edges occur on the same input) the circuit should stop counting while holding its last count.

7. Design a circuit with one input and two outputs that produces transitions alternately on the two outputs for every leading edge on the input. See Figure 3.13 for an example.

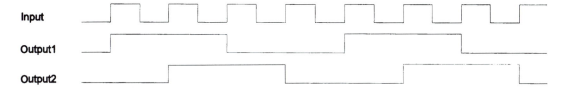

Figure 3.13 Quadrature Output, Problem 7

8. Examine the output from a mechanically actuated switch, looking particularly for its "bounce" characteristics. These usually appear as a series of transitions that appear immediately after the switch has been closed, resulting from mechanical bouncing of the contact. Design and implement a "debounce" circuit that will reject these artifacts.

9. Asynchronous input signals to synchronous circuits can cause problems when a transition of the input signal occurs too closely to the transition of the clock. Given a system with an asynchronous input (from a different signal generator than the one used to derive the clock) distribute the input to two D flip-flops. However, precede one of them with a pair of inverters to add some delay (Figure 3.14). Design a circuit with an integrated counter to detect and count the number of occurrences for which the flip-flops have different outputs, indicating a setup time violation caused by the asynchronous input signal.

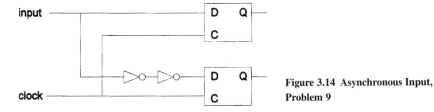

Figure 3.14 Asynchronous Input, Problem 9

4

Asynchronous Sequential Logic

It is not always possible or desirable to use synchronous logic circuits. Asynchronous circuits, which have no clock, can be faster because they don't have to wait for the clock and can be more direct in dealing with large numbers of asynchronous external elements. The standard synchronous components—JK, D, and T flip-flops—are themselves asynchronous circuits. On the other hand, asynchronous circuits are more difficult to design because the internal timing is not protected by the clock and can affect behavior. Asynchronous circuits are much less common than synchronous circuits.

4.1 FLOW TABLES

Because the inputs are always connected, transients must be accounted for in the design procedures. These transients are not important in synchronous design because the clock assures that all internal changes complete before any input changes occur. Flow tables are similar to next-state tables of synchronous circuits and are used to annotate the overall system behavior. Like transition logic, flow tables provide the connection between the engineering description of the system behavior and the circuit design. Unlike synchronous design, however, the flow table must account for unstable as well as stable states. Unstable states occur when an input has changed that will trigger a change to a new state but the associated state variables have not yet had time to change.

Each row in a flow table is a state, and the columns represent all possible input combinations. The table entries are next states. Stable states are shown with circles (or in boldface type) and unstable states are shown plain. An unstable state will move to another row before the transient completes. A duplicate set of columns is used for the outputs, which are associated with both states and inputs. In this case, associating outputs with states and inputs will make minimization easier.

The beginning of the design process is the production of a primitive flow table. It has only one stable state per row. Inputs are assumed to change one at a time, with the circuit allowed to reach a stable state before the next input change. This is a *fundamental mode circuit*. An alternative form of asynchronous circuit is a *pulse mode circuit*, for which the inputs must be short pulses, short enough so that the pulse is gone before any feedback signals arrive. Design of pulse mode circuits is similar to the design of synchronous circuits and will not be considered further here.

4.1.1 Set–Reset Flip-Flop Design

The SR flip-flop is itself an asynchronous circuit, so its design is an exercise in asynchronous circuit design. Given no previously defined memory elements, all memory will have to be constructed from feedback of outputs to the input. The primitive flow table for an SR flip-flop is shown below. To use the table, first find the present state of the system by finding one or more rows with the output matching the current output. Then find the column corresponding to the current input values. For example, if the current input is SR = 00, and the current output is 0, the corresponding row would be row a. Row a has a stable state for this position since its next state is also row a. If an input were to change, for example to SR = 01, stay in the same row but move over one column, to find that the next state is c. Move to row c, to find that this is a new stable state, with output still 0. Any number of changes can be tracked this way.

To construct such a table, start with the column headings as noted, and only a single row. Enter its stable state and the asssociated output, as shown.

State	SR 00	01	11	10	Output
a	a				0

Then take one possible input change, and enter a new state name in its column. Enter a new row for that state, and enter the stable state corresponding to the input change being considered. Enter its associated output as well, as shown.

State	SR 00	01	11	10	Output
a	a	a		b	0
b				b	1

Continue in this manner until all possible input changes have been considered. This completes the primitive flow table as follows:

State	SR 00	01	11	10	Output
a	**a**	c	–	b	0
b	d	–	–	**b**	1
c	a	**c**	–	–	0
d	**d**	c	–	b	1

The don't-care entries in the table correspond to transitions that are assumed not to occur. Since two inputs are assumed not to change exactly simultaneously, all of the transitions from stable states that require both inputs to change are marked as don't-care. Row c also has a don't-care for the 11 input, since that is not an allowed input for an SR flip-flop.

4.1.2 Flow Table Reduction

The primitive table is easy to visualize and construct, but it tends to have redundancy because of the restriction to one stable state per row. To combine rows, first check for rows with direct redundancy. These must have stable states in the same column, matching outputs, and compatible unstable states. There are none like that in this case. Next, look for rows that can be merged to put several stable states in a single row. Outputs must match, and corresponding entries must be compatible (similar to the implication table of synchronous reduction). In this case, a and c are compatible, as are b and d. Then rename the states and write a new (nonprimitive) flow table:

State	SR 00	01	11	10	Z
A	**A**	**A**	–	B	0
B	**B**	A	–	**B**	1

4.1.3 State Assignment

State assignment is not arbitrary for asynchronous systems. Incorrect assignment can lead to timing errors — *races*. This case needs only one state variable, so there is no problem in assigning values to it. The state variable is Q, with state A corresponding to $Q = 0$ and state B corresponding to $Q = 1$. A new flow table can be made with this state assignment:

	SR				Z
State	00	01	11	10	
0-	0	0	–	1	0
1	1	0	–	1	1

4.1.4 Excitation Map

The flow table gives information on next-state values, which are the values that will be fed back to provide the memory. The output of the flow table is the next state, which is the input to the memory element, just a wire in this case. The excitation map is shown in Figure 4.1. The excitation equation from this map is

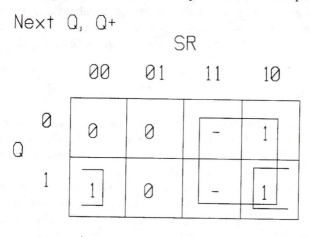

Figure 4.1 SR Flip-Flop Excitation Map

$$Q^+ = S + Q * R'$$

and the output equation is

$$Z = Q$$

The circuit is shown in Figure 4.2.

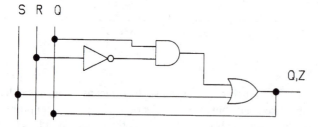

Figure 4.2 SR Flip-Flop Circuit

4.2 HAZARDS

There are three types of problems in asynchronous circuits:

- Logic hazards
- Races
- Essential hazards

Logic hazards are avoided by using methods described in Chapter 2 (covering terms). Races are avoided by proper state assignment, which prevents ambiguous state changes because of more than one state variable changing at the same time. This is discussed in more detail below. Essential hazards are more complex, but there is a procedure for detecting them.

The main method of detecting essential hazards is the *rule of three changes*. From any initial stable state, change the value of one input variable and observe the resulting stable state (S1). Start from the same initial stable state again, but make three changes rather than one to the same input variable. That is, instead of 0–1, use 0–1–0–1. Observe the resulting stable state (S3). If S1 is different from S3, a possible essential hazard exists. This is a timing inconsistency that can send the system to the wrong final state. The solution is to add appropriate delay to feedback signals, but analyzing the circuit to find out which delay is tedious. All possible input changes at all stable states must be examined. Simulation with accurate timing delays is an important tool here.

4.3 DESIGN OF A TOGGLE FLIP-FLOP

As with the SR flip-flop, the toggle flip-flop is designed as an asynchronous circuit. It has two inputs: toggle and clock. An edge-triggered flip-flop will be considered, with output changes assumed to take place on the rising edge of the clock. The primitive flow table follows.

Toggle Flip Flop

State	TC				Output			
	00	01	11	10	00	01	11	10
a	**a**	b	–	c	0	–	–	–
b	a	**b**	g	–	–	0	–	–
c	a	–	d	**c**	–	–	–	0
d	–	j	**d**	e	–	–	1	0
e	k	–	f	**e**	–	–	–	1
f	–	b	**f**	c	–	–	0	–
g	–	h	**g**	c	–	–	0	–
h	a	**h**	i	–	–	0	–	–
i	–	j	**i**	e	–	–	1	–
j	k	**j**	f	–	–	1	–	–
k	**k**	j	–	e	1	–	–	–

Note that the outputs have been made functions of inputs also (Mealy form). This makes reduction easier (the opposite of synchronous.). Rows can be matched more easily because of all of the don't cares. First, look for matching rows (i.e., directly redundant) with stable states in the same column, same output, and compatible unstable states. Rows d and i can be combined by these rules. Next, check for compatible rows for merging using an implication table (Figure 4.3).

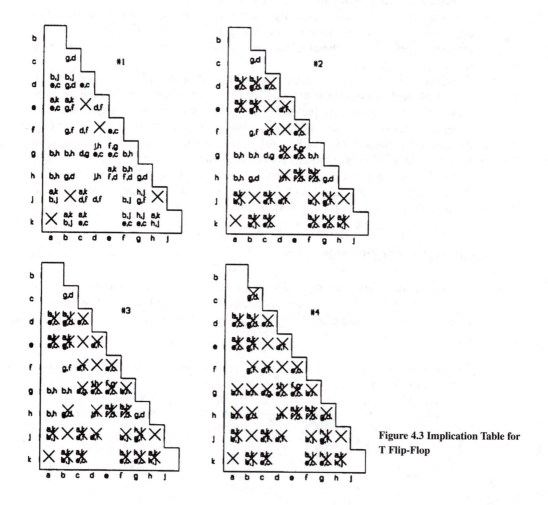

Figure 4.3 Implication Table for T Flip-Flop

From these procedures, the combinable rows are

a,b; a,c; a,f; c,h; d,k; e,j; e,k

Not all of these mergers can actually be made. There are no fixed rules for how to do the mergers. Try to pick the set that leads to the minimum number of rows, but for small systems, absolute optimization is not critical. To avoid confusion, rename the states A,B, . . . to get a new flow table:

	TC				Output			
	00	01	11	10	00	01	10	11
A (a,b)	**A**	**A**	F	B	0	0	–	–
B (c,h)	A	**B**	C	**B**	–	0	–	0
C (d,k)	**C**	D	**C**	D	1	–	1	–
D (e,j)	C	**D**	E	**D**	–	1	–	1
E (f)	–	A	**E**	B	–	–	0	–
F (g)	–	B	**F**	B	–	–	0	–

If two (or more) variables change "simultaneously," that is, faster than it takes for a circuit to reach a new stable state, the order of change will be interpreted arbitrarily. In a physical system, what this means is that random differences in otherwise similar components will govern which change takes place first. If those two variables are state variables, each interpretation will result in a different path through the flow table. (Input variables are "not allowed" to change simultaneously.) This is a *race* and can be critical or noncritical. Critical races end up at different final stable states, while noncritical races end up at same stable state. Since it is difficult to figure out whether a race will be critical or noncritical, state assignment is done to prevent races. This was not a problem in synchronous circuits because the clock prevented changes from becoming effective immediately.

4.4   STATE ADJACENCY

To avoid races, state variables must be assigned so that in getting from one stable state to another, only one bit changes at a time (i.e., only one state variable). This avoids both critical and noncritical races. To do this it is necesssary to identify all possible transitions from each stable state and then assign state variables so that no transition causes two or more state variables to change at the same time. For the example of the T-flip-flop, these are:

State	Adjacent states
A	B,E,F
B	A,C,E,F
C	B,D
D	C,E
E	B,D,A
F	A,B

A Karnaugh map can be used to assign state variables, since it is designed to deal with problems of adjacency. Put each state in a cell of the map and see if all of the adjacency conditions can be met. Figure 4.4 on the next page shows a trial configuration. It does not meet all of the adjacency requirements (open boxes are don't cares), nor does there seem to be any way to meet all of the adjacency conditions in this diagram. There are some open boxes, however. These can be used to enter *bridge* states, extra states to move from one of the original states to another without violating adjacency requirements. The bridge states are always unstable. For the T-flip-flop case, two extra states are needed (see Figure 4.5 on the next page). The new states, G and H, can be added to the state table, with the critical transitions routed through those states:

	TC				*Output*			
	00	*01*	*11*	*10*	*00*	*01*	*10*	*11*
A (a,b)	***A***	***A***	*F*	*B*	*0*	*0*	–	–
B (c,h)	*A*	***B***	*G*	***B***	–	*0*	–	*0*
C (d,k)	***C***	*D*	***C***	*D*	*1*	–	*1*	–
D (e,j)	*C*	***D***	*E*	***D***	–	*1*	–	*1*
E (f)	–	*A*	***E***	*G*	–	–	*0*	–
F (g)	–	*B*	***F***	*H*	–	–	*0*	–
G	–	–	*C*	*B*	–	–	–	–
H	–	–	–	*B*	–	–	–	–

Q3 \ Q1,Q2	00	01	11	10
0	A	B	C	E
1	F		D	

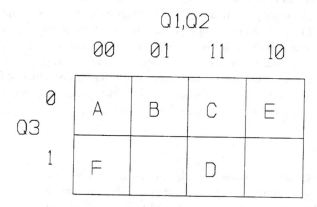

Figure 4.4 Trial–State Adjacency Map for T Flip-Flop

Q3 \ Q1,Q2	00	01	11	10
0	A	F	H	B
1	E	D	C	G

Figure 4.5 Adjacency Diagram with Bridge States

This gives the following state assignment for state variables Q1, Q2, and Q3:

A:000,F:010,H:110,B:100,E:001,D:011,C:111,G:101

From this, a flow table with actual state values can be generated:

	TC				Output			
	00	*01*	*11*	*10*	*00*	*01*	*10*	*11*
000	**000**	**000**	*010*	*100*	*0*	*0*	*–*	*–*
100	*000*	**100**	*101*	**100**	*–*	*0*	*–*	*0*
111	**111**	*011*	**111**	*011*	*1*	*–*	*1*	*–*
011	*111*	**011**	*001*	**011**	*–*	*1*	*–*	*1*
001	*–*	*000*	**001**	*101*	*–*	*–*	*0*	*–*
010	*–*	*100*	**010**	*110*	*–*	*–*	*0*	*–*
101	*–*	*–*	*111*	*100*	*–*	*–*	*–*	*–*
110	*–*	*–*	*–*	*100*	*–*	*–*	*–*	*–*

The procedure above might not have worked — there may have been too few open boxes to meet the adjacency requirements, even with bridge states added. The next step is to add additional state variables so as to increase the size of the adjacency map. In this case, the minimum number of state variables is three, so four can be used instead (Figure 4.6). The larger map will give extra possibilities for bridge states so that adjacency requirements can be satisfied. With enough state variables, it is always possible to assign race-free states.

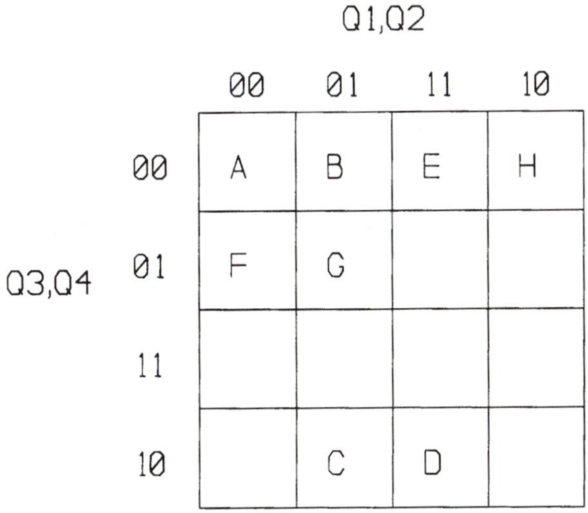

Figure 4.6 Augmented Adjacency Map

4.5 OUTPUT ASSIGNMENT

Output assignment is not a difficult step. Small changes when the output signals change are the only vagaries associated with output assignment. There are unassigned outputs associated with all of the don't-care entries in the flow table. When the unstable state is between two stable states with the same output, the unstable state must match that ouput. Otherwise, there will be an output glitch. For cases where the output is changing between the stable states, the unstable states can be set so that the change occurs at the beginning or end of the transition. They can also be left as don't cares so that the output circuitry can be minimized. An output assignment for the T flip-flop is shown below.

	TC 00	01	11	10	Output 00	01	10	11
000	**000**	**000**	010	100	0	0	0	0
100	000	**100**	101	**100**	0	0	–	0
111	**111**	011	**111**	011	1	1	1	1
011	111	**011**	001	**011**	1	1	–	1
001	–	000	**001**	101	–	0	0	0
010	–	100	**010**	110	–	0	0	0
101	–	–	111	100	–	–	–	0
110	–	–	–	100	–	–	–	–

4.6 EXCITATION NETWORK

Pick a memory element, which can be a delay or SR-flip-flop, then develop the excitation and output maps from the final flow diagram. Apply the three-change rule to see if there are any essential hazards; examine the timing if there are. Avoiding these associated hazards requires the insertion of additional delay elements, which is more complex than is warranted for this discussion. For large systems, simply applying the three-change rule is a lengthy process, and figuring out where and how much delay to apply is even more difficult. This is one of the factors that limits the practical size of asynchronous circuits.

4.7 ASYNCHRONOUS DESIGN FOR DOOR LOCK

The primitive flow table for an asynchronous design of the sequential door lock is given on the next page. The remaining design steps are left as an exercise. The combination for the lock is the same: press and release x1, then press and release x2. This generates the primitive flow table:

State					Next State x1,x2			Outputs x1,x2		
	00	01	11	10	00	01	11	10		
a	**a**	b	–	e	0	–	–	–
b	a	**b**	c	–	–	0	–	–
c	–	b	**c**	d	–	–	0	–
d	a	–	c	**d**	–	–	–	–
e	f	–	c	**e**	–	–	–	0
f	**f**	g	–	d	0	–	–	–
g	h	**g**	c	–	–	0	–	–
h	**h**	b	–	d	1	–	–	–

Note that some different decisions were made about what happens when the user enters an incorrect sequence. In the synchronous design, the system stayed in the same state; in this design it returns to the starting state each time an incorrect entry is made. This is more rational, since a sequence that includes incorrect entries must itself be incorrect. This could have been implemented in the synchronous design as well.

4.8 PLDS AND ASYNCHRONOUS LOGIC

PLDs provide for combinatorial output and feedback (i.e., no clock), and some also provide unclocked SR or JK flip-flops. The Intel software, however, does not provide for hazard-free minimized excitation equations. This can be done manually, but it is evident that the software is not intended for design of asynchronous systems. This is probably true of most design software since they are used most commonly for design of computer systems and computer peripherals. There is also no way to control delays in the circuitry other than using extra macro cells or external feedback. Thus PLDs can be used for the Boolean excitation circuitry but do not seem to be intended for complete system solutions as they are for synchronous systems.

4.9 PROBLEMS AND DISCUSSION TOPICS

1. Basic synchronous memory components are actually asynchronous circuits themselves, with the clock as one of the inputs. Design and implement circuits for the T and JKT flip-flops.
2. By controlling the order in which variables representing gate outputs are updated, simple computer programs can be used to simulate either synchronous or asynchronous logic. For example, for the logic fragment in Figure 4.7 the following code (in Matlab) represents a simulation of an asynchronous system for which each gate delay is 1 time unit.

```
% Input sequences
x = [1 1 1 1 1];
y = [1 0 0 0 0];
%Initial circuit values
```

```
v1(1) = ~x(1);
v2(1) = v1(1) | y(1);
v3(1) = x(1) | y(1);
v4(1) = v2(1) & v3(1);
t(1) = 0;

for i = 1:4
% Compute 'next' values
v1(i+1) = ~x(i);
v2(i+1) = v1(i) | y(i);
v3(i+1) = x(i) | y(i);
v4(i+1) = v2(i) & v3(i);
i = i+1;
t(i) = i-1;
end
% Separate values for plotting
v1p = v1+0.5;
v2p = v2+2;
v3p = v3+3.5;
v4p = v4+5;
yp = y+6.5;
xp = x+8;
[tt,v1s]=stairs(t,v1p);[tt,v2s]=stairs(t,v2p);[tt,v3s]=stairs(t,v3p);
[tt,v4s]=stairs(t,v4p);[tt,yy]=stairs(t,yp);[tt,xx]=stairs(t,xp);
plot(tt,v1s,'k',tt,v2s,'k',tt,v3s,'k',tt,v4s,'k',tt,yy,'k',tt,xx,'k');
%plot(t,v1p,'k',t,v2p,'k',t,v3p,'k',t,v4p,'k',t,yp,'k',t,xp,'k');
```

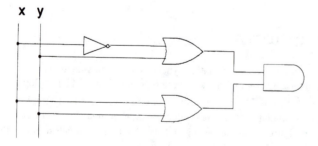

Figure 4.7 Asynchronous Circuit Logic Fragment

Figure 4.8 shows the response of the circuit to a change in the y input. Additional unit delays can be added by adding dummy gates. Use this method to simulate the circuits designed for Problem 1 and for other circuits in this chapter.

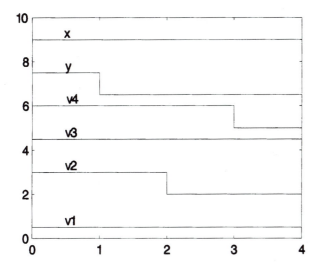

Figure 4.8 Asynchronous Circuit Response

3. Redo the problems of Chapter 3 using asynchronous circuits (no clocks).

5

Simple Computer Structure: Register Transfer Logic

When complex logic systems are designed it is necessary to modularize the design so that each module remains at a manageable size. Otherwise, the design procedures outlined in Chapter 4, even if implemented on computers and made more efficient, will bog down from the immense number of variables that will be generated. Debugging, testing, and validation also benefit from keeping module size modest, as does the ability to reuse modules in a number of products. Register/bus/processor design is a way to accomplish this modularization for systems such as general-purpose computers that need to transfer information as digital words, or groups of logic variables. Each module can then be designed independently. Because of the isolation afforded by synchronous design, integration should be relatively straightforward once the design of the overall system logic and the design of the individual modules is complete. Although errors in how the overall logic works are certainly possible, low-level errors such as timing of signals should not be a problem if the clock is run sufficiently slowly.

While it is unlikely that mechanical system design will often include full-scale computer architecture design, the material in this chapter serves two functions related to mechanical systems: (1) the modularization concepts can be used for custom interfaces; and (2) the general ideas on computer structure can be useful in the selection of industry standard computer systems for mechanical system control work.

5.1 REGISTERS

Registers are the basis for general processor design. They are used to hold information until it is needed; transfer from one register to another allows the processing to proceed. A register can be implemented as a set of D flip-flops, one flip-flop for each bit in the register. Figure 1 shows three bits of such a register. The purpose of the register is to be able to record information on command and to enable its three-state outputs on command so that the recorded value can be used by another component in the system (i.e., can be "connected" to the bus). The input recording is accomplished by the combination of the master clock and the load signal. As long as the load signal is off, the register will hold its old values because the clock will never get to the clock inputs of the D flip-flops. When the load signal is on, the next clock transition will cause the current input values to be recorded as the D states.

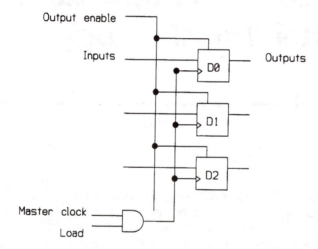

Figure 5.1 Register Made from D Flip-Flops

   The output is controlled by the output enable signal. When it is off, the output is in its high-impedance state, so will not affect any connected components. When it is on, the outputs are enabled (low impedance) and can drive a load.

5.2 DATA BUS

Registers are connected with a *bus*, a collection of wires in parallel. Each input wire of each register in the system is connected to one of the wires on the bus. Similarly, each output wire is connected to the same set of bus wires. Thus each wire of the data bus has both the input and output wires of one of the bits of every register connected to it (Figure 5.2). To avoid confusion, only one register at a time ever has its outputs enabled. All of the others will be in high-impedance state, so will not affect the bus. Therefore, the register that has its outputs enabled controls the information on the bus. Any of the other registers could read information from the bus. The only restriction there is that the output amplifiers be capable of producing enough current for all the potential inputs.

Data transfer on the bus is regulated by the load and output enable control signals. For example, to transfer data from register A to register B, the following sequence of control signals must be generated:

- *First clock tick*: set output enable on A to 1.
- *Second clock tick*: set load on B to 1; data are latched.
- *Third clock tick*: set output enable on A to 0, input enable on B to 0.

Practical master clock rates for systems of this sort are 10 to 50 MHz, or faster in specialized systems. The time for this operation would thus be 60 to 300 ns. With proper timing within the cycle, the operation could be accomplished with fewer clock ticks.

Additional logic units attached to the bus provide the needed processing power. Processors can be attached directly to the bus, to registers, or some combination. Figure 5.2 shows a possible configuration for an adder. This processor would add the contents of a dedicated register, treated as a binary number, to a value obtained from the bus. The result could be stored in any device connected to the bus; a likely spot would be back to the register from which the first operand was obtained.

The first step of the operation would be to get the first operand into the general register. Then the source of the second operand is identified (some other location on the bus) and the data are transferred from it to the input register of the adder. The source is then disconnected, and the general register's output is enabled as is the output of the processor's input register. The two inputs to the processor are now active and can produce its result, which is loaded into the output register. The final step is to enable the input of the general register and the output of the output register, thereby transferring the result into the general register, where it replaces the first operand.

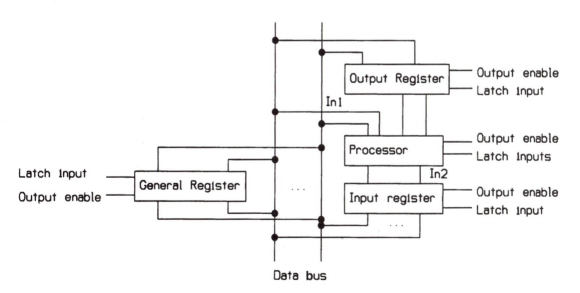

Figure 5.2 Bus, Registers, and Processor

5.3 CONTROL SIGNALS

The control signals in the previous section are not shown connected to anything, as if generated by magic. The actual source of these signals is the control unit, a logic circuit that is responsible for sequencing the control signals properly. All of the control signals are connected directly to the control unit; the bus is only for transmission of data. The complexity of the control unit depends on the task being performed. For a simple, specialized system, the control unit might be fairly simple. For a general-purpose computer, it gets quite complex.

This structure can be generalized to a computer CPU (central processing unit) which will contain:

- Memory access mechanism
- General registers
- Instruction counter (program counter)
- Memory address register
- Instruction register
- Stack pointer
- Arithmetic unit
- Input/output bus

Execution of a computer instruction involves a sequence of control signals that interact with these components. The sequence starts by using the instruction counter to find out where in memory the next instruction should come from. Then the memory is accessed and the instruction is brought from memory into the instruction register. The outputs of the instruction register are logic signals that are used to sequence the control unit. Depending on the instruction, the memory might again be accessed to get data, data might be moved among registers, an arithmetic operation might be performed, and so on. Each of these operations requires a different sequence of control signals. A set of all control signals is called a *control word*.

There are two common ways to implement the control unit. The most direct is to build it as a sequential synchronous circuit, using transition logic, synchronous design, and so on. The inputs to this sequential circuit are the signals from the instruction register, which govern the operation the computer is performing. The outputs are the control signals (the control word).

An alternative method for generating control words is to move the abstraction one level lower. The control word sequences associated with various instructions can be put in a memory—a RAM or ROM—but a fast one. The control unit is still a state machine, but it picks sequences out of the control word memory. The sequence of stored control words is called the computer's *microcode*. Note that the microcode memory is completely distinct from the computer's main memory, RAM and/or ROM. The microcode memory is much faster, has a much wider word (to accommodate the large number of control signals), and is much smaller in total size. The advantage of doing things this way is that the microcode is easy to change, so the processor can be customized for a variety of applications. The disadvantage is that it is generally not as fast as a dedicated logic circuit.

5.4 INPUT–OUTPUT BUS

The input–output (I/O) bus referred to previously is the connection between the CPU and the devices connected to the computer. The I/O bus was originally developed to connect the components needed to build a computer, several boards for the CPU, the memory, peripheral controllers, and so on. As component density in electronic devices has increased, this usage has given way to the current role for the I/O bus of connecting peripheral devices such as disk controllers, network adapters, and so on. I/O buses are now used less frequently for memory (RAM) because they are too slow.

Unlike the processor bus, described previously, which is strictly a data bus, the I/O bus contains all of the information needed to mediate data transfers, address, control, and data. These three types of information can be implemented on separate wires, which is the fastest way to do things, or some of the wires can be used for two different kinds of data, sequentially. The advantage of building a computer around a bus of this sort is that it is very easy to change the devices to which the computer is connected. This is useful for initial configuration, since it greatly reduces the amount of inventory needed and is also useful for future upgrades, even with equipment that wasn't invented at the time the computer was originally made.

To exchange information on the I/O bus, the master unit on the bus places an address on the bus. Each unit on the bus is responsible for a particular range of addresses. The relevant unit is alerted that it is expected to be part of a data interchange. It does this by constantly monitoring control signals on the bus, checking for those that affect it. Then the master unit indicates which direction the interchange will be. The appropriate unit puts the data on the bus, which the receiving unit reads, completing the interchange.

The bus protocol can be synchronous or asynchronous. That is, each of the units on the bus can be independent synchronous devices, or the units can all be connected by a master clock and act as a single synchronous unit. An asynchronous bus must use handshake signals to indicate when operations are complete, whereas in a synchronous bus, all operations must be completed within a clock cycle. The handshake slows the interchange down, and requires additional hardware, but allows for much more flexibility in designing the devices that will go on the bus.

Another complexity arises from the electrical nature of a bus, consisting as it does of parallel conductors. At any given time, only one device can be writing to the bus and is thus the *bus master*. In a *single-master bus*, one unit is the bus master, except when it voluntarily gives up the master status. In a standard computer, the master will be the CPU. It will give up its master status only in response to specific signals designed for that purpose. A good example of that is for direct memory access (DMA) devices, such as disk drives, which can read or write information directly to memory without CPU intervention. Otherwise, however, other devices can access the bus only when specifically "invited" to do so by the CPU.

It is often desirable, however, to allow more independence to the devices on the bus. This would allow, for example, for multiple computers to reside on the same bus. If several computers or other indpendent devices were to reside on the bus, they would need arbitrary access to the bus, and thus the bus would have multiple masters. This is called a *multimaster bus*. To implement a multimaster bus, circuitry must be added to the bus interface to adjudicate conflicts in bus access. The added complexity makes such a bus more expensive but much more capable than a single master bus. Multimaster buses are important in applications that have components that can act independently. If to accomplish their activity, these components need to make access to the bus, to use an analog-to-digital converter; for example, a multimaster bus is essential. In many cases, these "independent" components are implemented with a complete computer system built on a board that plugs into the bus. This is a major mechanisn for realizing multiple-computer systems where the computers must have high-speed communication among themselves.

5.5 COMMERCIALLY AVAILABLE BUSES

When bus structures were first developed for small computer systems, each computer manufacturer tended to develop a proprietary bus architecture. Since many of the components that could be used with these buses were made by third-party suppliers, it became evident that noncaptive bus structures would make this add-on market much more effective. This also became more relevant as computers shrank from multiboard to single board and even less. The need for close control of the bus structure by the computer manufacturer was thus greatly lessened.

The single most popular input–output bus is based on the structure used in the original IBM PC and, later, AT computers. This bus now goes by the name of *Industry Standard Architecture* (ISA). It is a single-master bus with a 16-bit-wide data path. Its major advantage is wide availability with large numbers of low-cost boards available for it. Its major disadvantages are its relatively slow speed, and the lack of strict standards for signal usage and timing. Bus setup is rather ad hoc, and conflicts among boards from disparate manufacturers are not uncommon.

This architecture has two direct successors, the *Micro-Channel Architecture* (MCA) and the *Extended Industry Standard Architecture* (EISA). Both of these are multimaster and both have 32-bit data paths. MCA was developed by IBM and EISA was developed by an industry consortium. Neither one has played a major role in mechanical system work, although both have the performance potential to do so.

There are also two buses specifically introduced as nonproprietary, generic buses: *Multibus* developed by Intel and VME bus developed by Motorola and others. Multibus exists in two forms, the original 16-bit version and a more recently developed 32-bit bus (*Multibus II*). All of these are multimaster buses and have been widely used in industrial control systems. The availability of a wide variety of single-board computer systems for them makes them ideal platforms for multicomputer applications. A number of similar buses exist, with roughly equivalent operating characteristics. They differ in terms of size, speed, data path width, and so on, so the designer can pick and chose to meet system requirements.

The speed requirements of high-resolution video has spawned yet another bus structure. It is a cross between the processor bus and the I/O bus. To establish a strong secondary market in bus-based components, bus structures have to remain stable over a time period that is considerably longer than the computer technology "half-life." Thus they are usually significantly slower than the CPU's raw speed. This may not be a problem for many peripheral devices, such as analog-to-digital converters, communication interfaces, and so on, but for video in particular, the data rate of the bus is a severe limitation. These new buses, commercially the *VESA* bus, developed by an industry association, and the *PCI* bus developed by Intel, try to meet these needs by acting as extensions of the processor bus. As such, they are also more limited in length, number of components that can be interconnected, and so on. In addition to video, these buses are beginning to be exploited for other high speed devices, such as disk drives.

5.6 MULTIPLE PROCESSOR BUSES

It is often useful to move several pieces of information within a processor at the same time. For example, there can be separate buses for instructions and data, or several buses for data, so that more information can be moved within the same number of clock cycles. Although system speed can be increased by increasing the speed of the master clock, this can be very expensive because all of the associated components must be able to keep up with the faster clock speed. Using multiple buses or other structural improvements allows lower-speed peripheral components to be used in systems with enhanced speed. The penalty is that the CPU must be more complicated, to control the operation of many buses simultaneously.

5.7 PROBLEMS AND DISCUSSION TOPICS

1. Connect several three-state devices to a common output line. Show that the device whose output-enable control signal is on determines the state of the output line.

2. Design a logic circuit to produce control signals to transfer data from one one-bit register to another. Extend the design to move an ON-bit around a circle of four one-bit registers. This design should include initialization circuitry that will guarantee that on power-up the first register is on and the rest are off.

3. Many computer lab I/O boards have a digital section with eight inputs and eight outputs. Using two bits of the output for address and as many additional signals as necessary for control, design a logic circuit that will enable one bit of the output to be directed to one of four one-bit registers.

4. It is often necessary to fill a register with successive "nibbles" from a bus that is narrower than the register. Design a control circuit that will successively write four output bits of a computer I/O board first to the low-order part of an eight-bit register, then to the high-order part. Use the other four output bits of the I/O board for control.

6

Embedded Control Computers

Embedded control computers are invisible laborers, hardly known to the general computing community. They are integral parts of mechanical systems of all sorts and provide the economics of cost and size that make "smart" machines possible. Modern integrated-circuit technology has made it possible to optimize computer chips in many ways, so embedded processors can be chosen to meet very specific mechanical system control needs. The term *embedded* refers to computers that are integral parts of other engineering systems. As such, they provide the computing power needed for operational decision making rather than the general-purpose computing applications common on the desktop. Embedded computers are in every manner of device: microwave ovens, automobile engine controls, avionics, laser printers, radio and television tuners, machine tool controllers (CNC), and others. The list is expanding all the time, so much so that it is becoming inconceivable to design almost any kind of powered device without at least considering computer control.

The enormous expansion in the range of applications reflects the rapid miniaturization still being achieved in microelectronics—so much so that computers for control applications run from under $1 up to whatever is needed to meet the computing demands of the application. Most of these embedded applications use computers that sport model numbers unknown to readers of popular computer magazines. Many of them are designed for embedded applications. The commercial success of PC-class computers, however, has also led to embedded versions of computers based on the 86 family of processors. Their familiarity to engineers, who use them for so many other tasks, is a big selling point, despite the fact that they were not originally designed for embedded use.

Although they are not very well known, they represent the most widely manufactured class of computer. Hundreds of millions of units are sold every year, in a variety of flavors. The most common types are:

- *Microcontrollers*: processors optimized for absolute minimum size by including virtually all functionality on a single chip (CPU, RAM, ROM, physical I/O, etc.).
- *Digital signal processors (DSPs)*: processors optimized for the ultimate speed in solving linear filtering equation. Widely used in communications applications; making inroads into control applications.
- *RISC processors*: either modified for embedded applications, or the same processors used for workstations. Used in applications that demand very high speed general computation in addition to monitoring and control of physical components. Controllers for laser printers are a popular application.
- *Embedded versions of the x86 architecture*: used because they are so familiar to everybody as a result of their use in PCs.

6.1 MOTIVATION, APPLICATION REQUIREMENTS, AND ARCHITECTURES

The needs of embedded control differ markedly from those of general-purpose microcomputers:

- They usually run a single program for the life of the machine in which they are embedded, with the exception of occasional field upgrades.
- They do not generally require mass storage or operating systems.
- Their user–operator interaction is very limited and prescribed; there might not be any user interaction at all.
- They may have to withstand hostile environments in terms of temperature, electrical noise, vibration, and contaminants.

Because of the singular nature of what they do, embedded computers can be optimized for their applications in ways that aren't possible for general-purpose processors because of the wide variety of tasks they must accomplish.

Although almost any computer could be used for embedded control, the discussion in this chapter focuses on applications in which the computer is physically as well as logically embedded into the mechanical system. These computers must meet the special needs of the embedded environment and are often customized to the application. Four computational architectures are discussed: microcontrollers, digital signal processors (DSPs), high-speed floating-point RISC, and integrated PCs, with the strongest emphasis on microcontrollers.

6.2 BASIC FEATURES OF EMBEDDED COMPUTERS

For high-volume cost-sensitive applications, embedded controllers must be, above all, compact and inexpensive. These factors have led to design decisions for processors targeted for embedded control that are quite different from the trade-offs driving the design of general-purpose processors. A single-chip computer has been a primary design goal for embedded control computers, in contrast to the single-chip CPU (central processing unit) that has been the focus of general-purpose microprocessor development. A single-chip computer means that problems requiring minimal resources can be solved with a computer system requiring only one chip plus a timing crystal. Even where single-chip integration cannot be achieved, the goal is still to minimize the chip count needed to implement a working system.

To achieve a stand-alone computer, the CPU, working memory, input–output, means for program loading, and, perhaps, a means for long-term data storage, must all be on the chip. The first problem is miniaturization. The initial breakthrough in embedded computer control systems came when all of these critical components could be fit on a single circuit board. Conventional processors were used for these single-board computers. The working memory was provided by static RAM and, later, by dynamic RAM also. The program was permanently stored in a ready-only memory (ROM). It could be executed directly from the ROM or copied into RAM for execution. Input and output were provided by parallel interfaces for communication with system components, serial interfaces for communication with terminals or other computers, and bus interfaces for communication with other circuit boards, if any.

6.2.1 Memory

Long-term, but not permanent data storage remains a vexing problem. Permanent data can be entered with the program and stored in ROM. Data that are generated during operation or entered by the operator, however, has no convenient storage medium. In conventional computer systems, magnetic disk or tape storage is used for this purpose, but they are not convenient for embedded systems because of cost, bulk, and susceptibility to environmental damage. Static or dynamic RAMs are solid-state devices. Static RAM retains information in flip-flop circuits, while capacitors are used for dynamic RAM (it is called dynamic because those capacitors that are charged must be refreshed periodically to avoid loss of information due to leakage across the capacitor). Both of these memory types require power to retain their information, so it cannot be used to store archival information. The ROM, on the other hand, cannot be written to, so it is equally useless for this purpose.

There are three general approaches to semipermanent data storage on embedded computers, none completely satisfactory:

1. *Battery backup.* Information stored in static RAM can be preserved if a battery is used to supply power when the main power is removed. The combination of CMOS circuits, which draw almost no power when they are not switching, and lithium batteries, which have a very long shelf life, can provide five to ten years of service before battery replacement is required.

2. *Nonvolatile, writable, solid-state memory.* Since the development of solid-state, read-only memory (ROM), extensive research has been directed toward relaxing the restrictions on writing to them without losing their nonvolatility (information does not "evaporate" when the power is turned off). The earliest (and still cheapest in large quantities) ROMs were mask programmed, that is, manufactured with the program in place. Field-programmable ROMs (PROMs) followed, which could be written to after being manufactured, but only once. The UVPROM can be erased by exposing it to ultraviolet light, then reprogrammed. The electrically erasable ROM (EEROM) is the only technology of this group suitable for semipermanent data storage. Its development is relatively recent, and it suffers several disadvantages with respect to other types of ROM and RAM. It is expensive and much less dense, requires long write times, and most important, can be written to reliably only a finite number of times. This technology is still undergoing rapid development. The *flash* form of electrically rewritable, nonvolatile memory is becoming the most common form of long-term storage. It is only rewritable in blocks, not word by word.

3. *Bubble memory.* This is a magnetic medium in which magnetic domains, bubbles, are moved in a solid-state substrate. Bubble memories are controlled in the same manner as disk memories. They are reliable and have no write-cycle restrictions. They are, however, bulky, slow, and expensive.

The single-board computer (SBC) ushered in a new era in machine control, but with further miniaturization, it became possible to shrink all of this to a single chip (except the nonvolatile memory). The skills gained in the application of single-board computers to control of mechanical systems were directly usable for the design of single-chip systems, but the reduction in both cost and size enormously broadened the number of problems for which computer control was appropriate.

6.3 ARCHITECTURES OF SINGLE-CHIP COMPUTERS: MICROCONTROLLERS

The first computers specifically designed for embedded work were the *microcontrollers*. They are single-chip designs with a broad enough price and performance range to be used in everything from consumer goods to industrial automation. In order to fit all the resources needed for a completely operational computer on a single chip, some compromises must be made. The usual design compromise is to use a much simpler main processing unit (CPU) than conventional microcomputers built using multiple chips. The payoff for the compromise is tremendous economy of size, since circuit features on the chip are about 1000 times smaller than equivalent features on a printed circuit card. Single-chip computers designed for embedded control applications, often called microcontrollers, are sold in vast quantities.

6.3.1 Basic Architecture

The earliest single-chip computers used the same feature set as the first single-board computers and were organized on the basis that the bulk of the on-chip memory would be needed for program storage, with only a small amount of read-write memory (RAM) provided for scratch data storage. Total on-chip memory was 1 to 4 kilobytes. The on-chip

RAM was often organized as an extended register set, so I/O features, regular registers, and data storage were all addressed similarly. Because they are still very economical to build and use, modern versions of microcontrollers in this class are still widely available. Substantially more powerful systems are now also available.

Although similar in concept to conventional CPU architectures, the single-chip computers were different enough to have unique instruction sets. In addition to unique instructions based on the simplified architecture, instructions were also added that were of specific utility for control. They thus required their own software development tools, which usually comprised little more than assemblers. Given the small amount of program memory available, use of compilers for development was impractical in any case. As more capable microcontrollers with more memory have become available, the software development environment has expanded as well, now including host-based (PC or workstation) C compilers, debuggers, and so on.

The on-chip memory normally occupies less than the full processor memory space, so external ROM or RAM can be used. To accomplish this, a simple bus is made available that can be used to expand the memory. This bus often shares its pins with parallel (digital) output ports; if external memory is used, some I/O capability is lost. A bus differs from a parallel port in that it is designed to support communication among an arbitrary number of devices. It does this by providing for address signals to indicate what device is to be addressed, control signals to indicate the operation(s) to be performed, and data signals for the information to be communicated. Devices attached to the bus need to have sufficient logic to know when they are addressed and to take appropriate action when they are addressed. The data section of a bus is inherently bidirectional since information can flow either to or from the CPU. Parallel ports, on the other hand, provide for direct connection to external signals. Each line of a parallel port is normally dedicated to a specific function, in contrast to the lines of a bus that serve multiple purposes. A parallel port can be either unidirectional or bidirectional, depending on the application.

The use of a bus brought out on external pins allows for attachment of any kind of device but requires a significant amount of external circuitry. This trade-off demonstrates a second major constraint in single-chip computer design. After restrictions due to the amount of circuitry that can reliably be put on a chip, the next most severe constraint is in how many I/O pins can be provided. The constraint is both physical, in having space for a large number of pins, and in manufacturing cost and complexity, since wires have to be connected from each pin to the interior of the chip.

6.3.2 Architectural Evolution

As chip densities increased, single-chip computers followed a development path similar to that of general-purpose systems: adding basic capabilities such as multiply/divide instructions, on-chip serial ports, larger memory space, and so on. With basic needs met, general-purpose processors have been moving toward the support of more complex operating environments requiring task switching among large, disk-based tasks, multiuser systems, and so on. Jobs done with single-chip computers have very different needs, however. If there is more than one task, the tasks are relatively short and are cooperative. External resources, such as disk drives, are virtually never used in single-chip, embedded applications, so the complicated memory management methods built into general-purpose CPUs would be wasted in single-chip applications. Instead of memory management and related areas, development in single-chip computers has concentrated on enhanced control-related I/O functions. Since small size and minimal assembly cost are primary reasons for using these microcontrollers, this goal can be optimized if as many relevant functions are brought onto the chip as possible.

6.3.3 On-Chip Input–Output

The most generic I/O function is the parallel port. Once the basics of a functional computer could be fit onto a single chip, parallel ports were next. Parallel ports can be either *unidirectional* (i.e., dedicated to either input or output) or *bidirectional* (i.e., capable of reading or writing data as the need arises). They can also be *dumb*, also called *transparent*, or can include a handshake protocol. Fully bidirectional ports must contain a handshake protocol so that they can react properly to external demands for read or write operation.

Mechanical system control most often needs the simplest of these configurations, the unidirectional, transparent port. Signals from the computer appear as voltages on the output lines on command; values of the input voltage can be read at any time. The signals coming into the computer could be from a proximity sensor, limit switch, and so on; the signals from the computer can control relay closures, motor on–off signals, and other kinds of on–off actuators. The parallel port is important for its generic use, but also because many other kinds of devices can be interfaced through a parallel port.

Timers and serial ports have about equal priority for the next features to be brought on-chip. Timers are used as part of measurement: for example, to determine the time between pulses coming from a mechanical device. They can also be used to synchronize events, such as constructing a pulse-width modulated (PWM) signal. Internal timers are usually 16-bit, with a 1-or 2-μs count rate; often, several are available.

Serial ports provide a means for microcontrollers to communicate with other microcontrollers, with host computers, or with an operator. A serial port is important for this task because of the wide standardization for the electrical (RS232 or RS422) and the logical (ASCII) protocols, because software to drive the serial port is easy to write, and because only a few wires are needed, so long-distance communication is easy (using modems, the communication distance can stretch to anywhere a telephone is). There is often, however, a need for additional electrical interfacing to match the voltages available from microcontrollers (often, 0 to 5V) to the RS232 standard (12V).

Standard serial ports receive data in eight-bit packets (seven data bits plus one bit for error checking); an enhancement to serial ports that is often used in single-chip systems is the addition of a ninth bit. It is used to implement simple networks. In one state, the ninth bit indicates the transmission of a normal data packet. In its other state, the bit indicates that the data portion of the packet contains an address. This address can be used to pass a token to a particular processor, indicating that it can transmit information to the network. This simple multidrop network can form an effective coordinated system as long as its modest data rates are adequate for the job.

With parallel and serial ports, and timers provided for, analog-to-digital converters are a logical next step. This is, however, a difficult step because hybrid circuitry (analog and digital on the same chip) is much more complex to design than is purely digital circuitry. Microcontrollers that provide A/D conversion on the chip generally provide multiplexed, successive-approximation converters with 8 to 10 bits of resolution. Digital-to-analog converters are less common, but pulse-width modulation (PWM) is frequently built into some single-chip computers.

6.3.4 High-Speed Input–Output

High-speed input–output facilities are of particular usefulness for mechanical system control and are included in some microcontrollers. They are specialized for handling events that are characterized by transitions in digital logic signals. In particular, they provide a means of handling these events with much more precise timing than is available with software or interrupts, as long as the frequency of the events does not get too high.

The input side does this by adding a time stamp to the basic event capture, and storing the event information in a buffer. When the specified transition takes place the fact that it took place, and the time at which it took place are both recorded. This information can be accessed by software, either through polling or interrupt. In either case, though, the time stamp gives information on when the event actually happened, with a resolution of 1 or 2 us rather than when the software detected that it happened. This gives at least an order-of-magnitude improvement or more in the time resolution on events. This is valuable, for example, if a rotating shaft is generating pulses on a regular basis. Velocity can be determined from the time duration between successive pulses, which is obtained from the time stamps.

The event-frequency restriction is governed by the speed at which the software can remove events from the buffer and rearm the high-speed input for the next event. This is normally much longer than the time resolution of the high-speed input. On the other extreme, the longest event that can be captured is governed by the width of the clock/counter. Typical maximum times are 50 to 100 ms.

The output side is complementary—it creates events at hardware time resolution. It uses a content-addressable memory to store the information about the event to be created. Software enters information about the specific event (which output signal) and the time at which the transition is to occur. The high-speed output controller continually monitors its memory and executes the desired transitions when the specified time is reached. As with the high-speed input, the frequency is limited by the rate at which the software can put fresh events into the content-addressable memory; the maximum time until the event is to occur is limited by the maximum count of the clock. A free-running (i.e., nonresetable) clock is normally used for these modes to avoid timing ambiguities among events.

6.3.5 Microcontroller Configurations

Microcontrollers are available with 4-, 8-, and 16-bit internal data paths. Unlike the general-purpose computer market, where narrower data widths invariably mean old and undesirable, the microcontroller market is so price sensitive that making the computer fit the application is essential. Because of this specialization, the feature sets that are available are widely varied. They would include the type of on-chip memory, assortment of I/O interfaces included, chip speed, packaging type, and so on. On-chip memory almost always includes some amount of scratchpad RAM, but might include masked ROM, field programmable ROM, erasable ROM, or no ROM.

6.4 DIGITAL SIGNAL PROCESSORS

DSPs come from very different technological needs than those that spawned microcontrollers. They have been driven by the needs of communication systems, where filtering sampled data streams is the primary task. In voice systems, for example, it is constantly necessary to remove unwanted spectral components from signals in order to maintain the integrity of the information. Digital filters offer many advantages over analog filters in terms of their ease of design, component stability, and ability to implement highly tuned filters. Early microprocessors, however, were incapable of meeting the processing speeds needed, even for the telephone network, which only requires a bandwidth up to about 3 kHz. The structure of DSPs evolved directly from the form of digital filtering equations. Finite-impulse response filters have equations of the form

$$y_k = \sum_{i=0}^{n} K_i u_{k-i} \qquad (6.1)$$

where k refers to the kth sample in the data stream, the K's are coefficients, the u's are current and past values of the input, and y is the filter output. An impulse input is a signal for which the input has a nonzero value for one input sample, and is zero otherwise. This is called a finite-impulse response filter (FIR) because the output will stop changing n time steps after the impulse has entered the filter.

An infinite-impulse response filter (IIR) has an equation of the form

$$y_k = \sum_{i=0}^{n} K_i u_{k-i} + \sum_{i=1}^{n} C_i y_{k-i} \qquad (6.2)$$

where the Cs are another set of coefficients.

Because the output values are "recycled," the response can keep changing long after the input has stopped changing. For example, for the very simple filter

$$y_k = u_k + 0.5 y_{k-1} \qquad (6.3)$$

the unit impulse response (i.e., $u_0 = 1$, all other u's = 0) is the sequence, 1, $1/2$, $1/4$, $1/8$, ...

There are advantages to each of these forms, but from a computing perspective, they both take on the form of weighted sums. DSPs were developed to solve these filter equations quickly. By incorporating a multiply-and-accumulate unit (MAC) and bus structures that can get coefficients and input and output values efficiently, DSPs could achieve computational speeds for filtering equations many times faster than general-purpose processors. The rest of the architecture was designed to provide rudimentary input and output so that the signals can be brought into the DSP and the results sent out, and the necessary instruction set for the usual housekeeping chores that must be done. The result is a chip designed specifically for embedded applications in which filtering is the major task.

Other application areas have discovered that DSPs can be used profitably. Feedback control is one of these, since the linear feedback equations are expressible in exactly the same form as the filter equations given previously. These have been used in applications requiring unusually high sample rates, such as head positioning control on disk drives. Originally, DSPs supported only fixed-point calculations, so very careful scaling had to be done to make sure that there were no under- or overflows in the computation. Recent advances in circuit miniaturization have enabled the production of DSPs that can work directly in floating point, reducing the design load on the programmer.

6.5 RISC AND HIGH-SPEED FLOATING POINT

A problem that has faced many embedded applications is the lack of good floating-point support. Microcontrollers have been optimized for input–output and on-chip memory, so thus far there has not been room for integrated floating point. DSPs now have floating point but have a more limited instruction set than do standard processors. RISC processors have filled that gap in a number of embedded applications. In the high-volume arena, for example, controllers for laser printers need to do complex geometric calculations to position dots on a page and yet are part of a very price-sensitive market.

The reduced instruction set computer (RISC) is a design strategy for processors based on simplifying the CPU architecture so that most instructions operate in a single CPU cycle, there are a large number of general-purpose registers, and all instructions except load and store operate only on registers. This is in contrast to the (then) conventional design in which the instruction set included a wide variety of addressing modes and the ability to complete complex operations within a single instruction. With the advent of RISC, these became CISC (complex instruction set computers). RISC designs are premised on the use of compilers for the bulk of programming. The instruction set does not lend itself well to assembly language programming, whereas CISC instruction sets are often optimized for assembly language programming.

RISC processors, being relatively simple, use less of the available space on the chip for logic than do equivalent CISC processors. One of the uses to which that extra space was put was for floating-point processors. Although most of the RISC designs were targeted for workstation applications, some were designed specifically for embedded applications. In addition, some designs that were not widely accepted in the workstation market were retargeted for the embedded market.

The embedded applications for which RISC processors are used are computationally intensive and tend to require substantial amounts of memory. There is no possibility of achieving a single-chip design in this case, so embedded RISC processors are not that different in their configuration from workstation RISC processors. The main difference is that the needs for extensive memory management facilities are much less in embedded applications. Once a multichip solution is necessary, the input and output can also be off-chip so that it can be customized for the application.

6.6 EMBEDDED APPLICATIONS OF PCS

The IBM-PC standard was rapidly adopted for laboratory computing, including data acquisition and control. The hardware design included an open architecture bus, so specialized input–output interfaces could be added, and large numbers of compilers became available for software development. PCs configured for embedded control were a natural outgrowth of the widespread lab use of PCs, but they had neither the computing power of RISC processors nor the compactness of microcontrollers. This situation is now changing, for two reasons: (1) the development of PC processors with on-chip floating point, and (2) the miniaturization needed for portable computing. With these features, PC-based designs fall on a continuum between microcontrollers at one end and RISC systems at the other end. That, plus the familiarity engineers have with them, makes them a very attractive solution because design and implementation time can be minimized.

6.7 WATCHDOG TIMERS

Watchdog timers are common safety devices that are often included as on-chip facilities of microcontrollers or in external circuitry for other embedded applications. Their function is to detect and, if possible, recover from software failures. They automate the process so familiar to users of desktop computers—if the computer goes "dead," as evidenced by no activity for a suitably long period of time, reboot or reset the computer and start again. If the failure was truly software, the reset process might be able to bring the computer back to its previous state and allow work to continue. There may have been some loss of data which the user is responsible for recreating.

Embedded computers often have no operator nearby to perform a reset and may not even have an operator interface, so the reset function has to be autonomous. The general scheme of a watchdog is that a timer is set and started. It is the responsibility of the software to reset the timer to its original value periodically. If the timer ever runs down to zero, the watchdog circuitry will cause the computer to execute a hardware reset, allowing the control program to reinitialize and try again.

Design of the watchdog-related software is aimed at:

• Avoiding false alarms (i.e., watchdog-generated resets) when there is really nothing wrong with the software
• Assuring that triggering will actually take place if the software malfunctions

The first of these, avoiding false alarms, is the easier of the two. It requires careful analysis of the control code to identify the longest possible time between watchdog checks so that the watchdog's timer can be set realistically. The second is much more complicated, since an objective definition of correct behavior must be available before incorrect behavior can be reliably identified. For single-thread synchronous programs (no interrupts, no preemptive multitasking), an arrival condition is usually deemed an adequate approximation of correctness. Watchdog checks are placed at strategic points in the code. If the program actually arrives at the checkpoint before the watchdog timer runs out, it simply reinitializes the watchdog. Software for this method is simple — since "correctness" is defined as reaching the checkpoint, the software only has to reinitialize the timer.

When multithread software is used, however, the watchdog software can no longer be so simple. To take a particularly egregious example, imagine a single-thread program (as described previously) in which an interrupt is added whose only function is to reinitialize the watchdog timer periodically. This scheme will not catch any software failures in the actual control program unless the failure also interferes with the interrupt system. Even if the control program stops executing entirely (by getting into a small infinite loop, for example) the interrupt will continue reinitializing the watchdog timer so that no error is detected. This is a gross violation of the second design criterion listed previously.

This situation could be corrected if the control program generated state information for the watchdog software or watchdog monitor. Each time the watchdog monitor runs, it would check the state information from the control program to see that its execution is proceeding correctly. The exact equivalent of the single-thread system could be reproduced if the state information that is generated is arrival information at watchdog checkpoints. This general scheme, however, is not limited to checking just a single task. Any number of tasks could be checked in this manner; incorrect state information from any task would trigger an error condition, as would a failure that prevented the watchdog monitor itself from running.

Even this solution has a "gotcha," however. In general, there is no way to tell how much CPU access any particular task in a multitasking system will get since task preemption is governed by external events. There are several solutions to this problem, all of which make the watchdog monitor more complicated. One is to establish critical times for each task such that they must meet certain stated progress within specified time intervals. This would establish minimal timing constraints as part of the error structure. Tasks that only run when asynchronous external events occur, however, could not live up to this standard. To cover that case, the monitor would have to know whether a task is supposed to be running, and only check it if it is. To do that effectively, the monitor would probably have to be part of the preemptive scheduler.

These discussions could go on forever, examining ever deeper correctness issues, but the idea remains the same—define correct behavior and provide a mechanism to check for correctness. The beauty of the watchdog method is that it will cause an error condition even if the error-checking software (the watchdog monitor) itself fails.

6.8 SOFTWARE DEVELOPMENT ENVIRONMENT

Developing software for embedded systems has two components that distiguish it from development of "conventional" software:

- Embedded software is real time and includes asynchronous external and internal components.
- The target computer is often different from the development computer and is likely not to have the video and storage facilities that are so essential to debugging.

The tools available to remedy these problems range from very little to very extensive and can make a major difference in the efficiency with which a project is completed. Hypothetical examples at several of the extreme points can illustrate. The dimensions to be considered are the nature of the application (whether the application itself requires high-speed video, disk storage, etc.), what kind of programming is needed (assembly, compiler, etc.), economics of development cost versus manufacturing cost, and computational complexity of the application.

6.8.1 Example 1: Consumer Product, Four-Bit Processor

This application has to service a pushbutton/numeric display user interface, its control functions are entirely on–off, and the time constraints are not difficult to meet. Given the constraints of cost, which limit the amount of memory that can be used, and the CPU capabilities, assembly language is indicated for the programming. With no mass storage in the production product, the program will have to be stored in read-only memory (ROM). The device does not keep any long-term or archival information, so no external storage or printing is needed. Examples of the products that this specification might apply to include microwave ovens, radio tuners, VCRs, and home thermostats. Some of these, for example the home thermostat and the VCR, might need some local nonvolatile memory to store user-generated program information.

In the minimal case, the tools available would be a cross-assembler and a PROM burner (programmable read-only memory). The cross-assembler would run on a desktop computer and produce code for the target machine. The output from the cross assembler would go to the PROM configuration software and then to the PROM burner, where a PROM would be "burned." The PROM is inserted into the target computer, and it is tested with the prototype (or actual) product.

The production product would undoubtedly use mask-programmed ROM, and all the chips would be soldered in place since the only field solution to a bad computer would be to exchange the entire board. To develop code using the foregoing scenario, the production system would have to be modified to use a socket for a PROM that is pin compatible with the production ROM. The problems with using this method for development are that the turnaround cycle for a program change is very long (a PROM must be erased, burned, and inserted for each change) and the only debugging information available is from the numeric display. The advantage is that the cost for development tools is very low.

6.8.2 Example 2: Technologically Complex Control Application

This is still a microcontroller application, but the control involves both digital and analog information, and the real-time constraints include several interacting tasks. A high-end 8-bit microcontroller or a 16-bit version is called for to meet the computing needs. Examples of this type of application would include simple automated machinery, control of a unit process, control of a robot axis, and distributed energy management control. There might or might not be a user interface; if one were to exist, it would be of the pushbutton/numeric display type.

The application complexity and more relaxed cost constraints would favor using a compiler language (C, Pascal, etc.) for most of the programming, with assembly, if any, reserved for time-critical portions. If product volume were reasonably high, a totally assembly language version would be possible. The high volume would presumably amortize the added cost of assembly language programming as against the unit-cost savings of being able to use a less capable processor and/or less memory.

The simple solution outlined above would almost certainly be unsatisfactory. The computing complexity is high enough that much more debugging information would need to be available, and the number of code turnarounds would make the PROM process very inefficient. A high-end solution would be to use an in-circuit emulator (ICE). These devices plug into the target computer board in the same way as the microcontroller, but include access to the inner workings of the processor as well as RAM emulating the ROM that is used on the target system.

Developing software with an ICE is almost like developing software directly on the target platform. After compilation and/or assembly, the code is loaded into the ICE's memory, which is emulating the target system's memory. This process is essentially the same as loading a program to run on a desktop computer. The debugging software has access to the ICE's innards, so debugging the program is also similar to debugging a native mode program. Breakpoints can be set, registers can be examined, and so on.

An intermediate solution, usually considerably less expensive than a full ICE, is to design a prototype version of the control computer with sufficient additional equipment to facilitate debugging and testing. For example, a reasonably high-speed link between the host computer and the target could be used for downloading code and for sending back debugging information. This requires a monitor program running in the target computer and sufficient RAM to load the complete program into. If the cost of memory is not excessive with respect to overall system cost, the monitor and RAM can be left in place in the production system and used for field diagnostics and upgrades. The additional communication interface, memory, and so on, would not be present in the production version.

6.8.3 Example 3: Computationally Intensive Signal Processing or Control

These types of applications have mathematically complex algorithms to be implemented, usually requiring computing speeds that are a stretch for embedded technology. These applications include control of flexible structures, processing of vision information, and acoustic processing. Microcontrollers do not have sufficient computing power for these applications, so a DSP, RISC, or general-purpose processor will have to be used.

The development problem in these sorts of applications is often the processing algorithm itself: that is, debugging the algorithm, trying it on data from prototype systems, checking for convergence rates, instabilities, and other mathematically related issues. A common solution to this portion of the development cycle is to isolate it by using plug-in boards containing the target processors. These boards go onto the host computer's I/O bus, which could be a PC (ISA, EISA, or MCA), VME, Multibus, or other common bus structure. By operating the processor on the host's bus, its use becomes only slightly more complex than use of the host computer itself. This provides an effective environment for generating and analyzing data, making quick program changes, and so on. The remainder of the development process can proceed as described in the previous example.

6.8.4 Example 4: Applications with Database or Archival Needs

Systems with database or archival needs require a mass storage subsystem, usually a disk drive. With the addition of a disk drive, the target system can run a full operating system, and so can be its own host. With host and target the same system, software development proceeds exactly as it does for other native applications. Examples of these applications include manufacturing applications requiring large amounts of process information; process applications such as pharmaceuticals, that have extensive archival requirements; and systems that generate large amounts of information, such as test aircraft data acquisition.

6.9 PROBLEMS AND DISCUSSION TOPICS

These problems deal with situations in which microcontrollers solve problems that are typical of simple machine applications. There are many additional applications associated with later chapters, which will be dealt with as they come up. The problems refer to using microcontrollers for most of these applications. If microcontrollers are not available, a standard PC can also be used if it is equipped with a laboratory interface board.

1. In their early days, a major application that was predicted for microprocessors was as a replacement for hardwired logic. Given the limited capability of these early processors, it was not unreasonable that ambitions did not go much further than logic. Even today, many applications of the smaller microcontrollers can be viewed as applications of sequential logic. In that sense, most of the logic problems given in preceding chapters can be solved with microcontrollers. Solve a selection of those problems. Use direct simulation of the logic equations (i.e., write out the logic expressions directly in whatever language is being used) and also use if-then-else and computer language–based programming to solve the problems. Measure relative operating speeds of the computer and the hardwired logic circuits and comment on relative ease of design, setup, and debugging and testing.

2. Optical inputs are very common in manufacturing or other applications where large numbers of parts are handled. These can be either occlusion (beam breaking) or reflective devices. A common application is to measure the size of objects that are sitting in front of a set of such optical detectors.

    (a) Set up and test a circuit that can be used to bring a signal into a microcontroller from an occlusion type of sensor.
    (b) Use several such sensors and write a program to use them to detect the length of an object. Check for error conditions as well.
    (c) Use a single sensor and a moving belt to measure the length of an object moving by the switch at constant velocity.

3. Fixed numerical displays are one of the simplest forms of operator interface. The seven-segment LED (or LCD) display has enough resolution to (crudely) display all the numbers and some letters as well (Figure 6.1). Each segment is connected to a separate logic line which turns that segment either on or off. Combine such a display element with the length-measuring program to display the length measured or an "E" for error.

4. Communication with the outside world is often a problem with microcontrollers. Using a microcontroller with a buit-in serial port, design simple communication software that can send a multidigit integer value to a desktop computer for display.

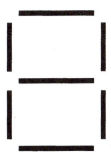

Figure 6.1 Seven-Segment Display

5. Many microcontrollers have built-in analog-to-digital converters. Use the converter to read in a value from an electronic temperature sensor.

   (a) Using an external thermometer, calibrate the sensor and determine its accuracy, precision, and noise level. Is the calibration linear? If not, try constructing a nonlinear compensation.
   (b) Construct software to take sequential readings from the thermometer and then use the serial port software from part (a) to send the result to a desktop computer for display.

6. Make a list of small appliances that might use microcontrollers. Pick several representative samples and design hardware and software that would be needed to reproduce the observed behavior. Build a simulation of the system using common laboratory components.

7

Stepping Motors

Stepping motors can provide accurate positioning without the need for position feedback instruments. This property makes them very attractive for a variety of applications, particularly those for which the cost of the instrument and its associated electronics are a significant proportion of the system cost. Other reasons for using stepping motors in an application include cases where the operating environment might make it more difficult to use a sensor, or situations in which tuning a servocontroller is difficult or inconvenient. Stepper motors use no brushes, so are also relatively low maintenance.

The stepping motor is an ac synchronous motor, designed to be operated using digital excitation for each of its windings. Because constant voltages are used when the motor is holding a position, they are sometimes called dc stepping motors. In a synchronous motor, the motion of the motor follows the phasor motion of the electrical excitation. In normal cases, the synchronous motor will track a speed governed by the frequency of the electrical driving signal, as in an ac electric (analog) clock. The term *synchronous motor* usually implies smooth, continuous motion. In the case of the stepping motor, however, the principle of following the phasor of the electric driving signal still holds, but the input signal is now digital and can hold the same phasor angle for long periods of time. During these times, the stepping motor will hold a position, applying an opposing torque if any torques are present that try to move it away from that position.

When the electric drive signal changes, it makes a sudden jump to a new phasor angle, and the motor tries to follow it. If the motor is not overloaded, either by static load or inertial load, it will move to that new position and then hold the new position until the excitation changes again. If the period of change of the excitation is kept constant, the motor will run at a constant average speed. However, the speed within each step could change by a considerable amount. It is thus very difficult to run a stepping motor in true constant-speed operation, although it is very easy to run it at a constant stepping rate. Other disadvantages include vibration in the mechanical system induced by the uneven motion of the motor, high electric current usage when the motor is stationary, and high temperatures. If the physical system has a lightly damped resonance, there may be *critical speeds*, speeds that the motor should not be run at because of the possibility of damaging the physical system. Stepper motors come in two major types: permanent magnet and variable reluctance. In general, the permanent-magnet motors perform at a higher level and are more expensive.

7.1 PERMANENT-MAGNET STEPPING MOTORS

The most common kind of stepping motors use permanent-magnet rotors. These are toothed rotors, built so that the points of the teeth are all of the same magnetic polarity. The stator consists of wound pole pieces, usually with double sets of windings on each pole piece (bifilar wound). Depending on which winding is excited, a pole can become a magnetic north or south. For any excitation pattern of the poles, the rotor will have a stable equilibrium position that it will seek. While in that position, any pertubation that moves the rotor away from equilibrium will cause a magnetic torque on the rotor that will tend to return it to the equilibrium position. This is the *holding* torque of the motor and establishes the level of static load that the motor can withstand without losing its position.

If an excessive load is applied, the rotor will move far enough away from the equilibrium point to be attracted by another equilibrium point, and if the load is relieved, will settle at that point. If this happens, the motor has lost its commanded position, but since there is no feedback, the control computer has no way of knowing that the position has changed, so a permanent error is made.

7.2 STEPPING SEQUENCE

Figure 7.1a shows an idealized stepping motor with a three-toothed rotor. For illustration purposes the rotor is shown as a triangle. It is in a stable equilibrium position. Because the magnetic forces are distance dependent, the closest pole-tooth sets will govern the resulting motion for small displacements away from the equilibrium. In the configuration shown in (a), small motions will generate a restoring torque. To cause the motor to take a step, the excitation must be changed in such a way that the resulting torque on the rotor will move it in the direction of the nearest stable equilibrium. There must not be any unstable equilibria between the rotor's current position and its intended position so that the torque on the rotor will continually move the rotor toward its next stopping point.

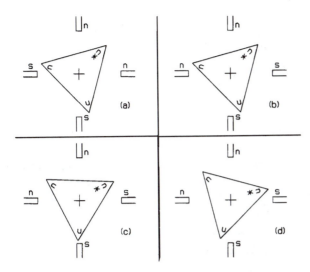

Figure 7.1 One-Step Motion for a Stepping Motor

Figure 7.1b shows the stepping motor immediately after the electrical excitation has been changed. The change in excitation consists of exchanging north and south on the horizontal poles, either by reversing the current flow through those coils, or by energizing the alternate set of coils on each pole. The rotor is now not in an equilibrium position; the torque on the motor will accelerate it in the clockwise direction. Figure 7.1c shows the rotor in motion, on its way to the new stopping point (the * marks one tooth on the rotor so that its motion can be visualized more easily), and Figure 7.1d shows it at its new equilibrium. This procedure results in the motor completing one (full) step.

The next step is taken by reversing the magnetic polarity on the top and bottom poles. When that happens, the next equilibrium position will be with the starred tooth at an angle of 15° below the horizontal, for another 30° step (it started 15° above the horizontal). This motor will thus produce 12 steps per revolution. If instead of reversing the excitation on the top and bottom poles, the excitation on the left and right poles had been put back to its original state, the motor would have stepped back to its initial position. The logic for stepping the motor is alternately to reverse the excitation on the pole sets. It takes four such reversals to complete a cycle, which is then repeated.

7.3 EXCITATION SEQUENCE

This procedure can be expressed as a logic process. For a double (bifilar)-wound motor, there are two coils on each pole, with the coils on opposing sets of poles coupled. The poles are divided into two sets, horizontal and vertical for a simple motor. A commercial motor will have more poles and teeth than this to achieve more steps per revolution, but the excitation wiring has the same arrangement externally, and the excitation sequences are the same. A motor with more poles and teeth will just take a smaller step for a single change in excitation. Each set of windings is wired together to give four inputs, two sets of poles times two windings, which we will denote as A, B, C, and D. A and B apply to one set of poles, C and D to the other. Exciting the A input excites one of the bifilar windings, B the other. Since the pairs are oppositely wound, A and B must never be excited simultaneously, and C and D must also never be excited simultaneously.

Table 7.1 shows the magnetic pole positions as a result of energizing each of the coil pairs. Changing excitation by proceeding down through the table causes motion in one direction (clockwise in the example in Figure 7.1); going up in the table causes motion in the opposite direction. The ends of the table wrap around; that is, when the bottom is reached, the next excitation value is taken from the top. Forward progression through this sequence gives motion in one direction, reverse progression gives motion in the other direction. The sequences wrap around at top and bottom to give continuous motion. This four-phase excitation sequence is very often used for permanent-magnet stepping motors, however, other sequences are possible and might be necessary for motors with poles that are wound differently. The four-phase sequence produces the highest available torque.

TABLE 7.1 FULL-STEP EXCITATION SEQUENCE

A	B	C	D
1	0	1	0
0	1	1	0
0	1	0	1
1	0	0	1

AB	Right/left	CD	Top/bottom
10	NS	10	NS
01	SN	01	SN

7.4 PHYSICAL CONFIGURATION

To make a permanent-magnet rotor all of whose teeth have the same polarity, the permanent magnets can be laid out axially. Each longitudinal section of the rotor thus has the same magnetic polarity all around its periphery. For adjacent sections, the stator windings are simply reversed, so that the effect on the torque is the same. A very common configuration for permanent-magnet stepper motors is four-phase excitation with 50 teeth, giving a motor with 1.8° steps, 200 steps per revolution.

7.5 PULSE EXCITATION

A common mode of operating stepping motors is to provide an interface that generates the four-phase excitation from pulse and direction inputs. The direction input is a level signal that controls the direction in which the motor is to step. The pulse input causes the motor to take one step in the specified direction for every pulse. This is a very convenient and intuitive way to drive a stepping motor. The required pulse trains are easy to generate either from software or from a logic system. Only two bits are required for the interface, and the drive logic becomes independent of the excitation sequences needed for specific motors.

Figure 7.2 shows a pulse-direction controller as viewed by the user. The system generating commands for the motor only has to pulse the input line to cause the motor to step. The direction input is a logic level; the direction is the value of that level at the time the pulse input goes through a transition. A simple way to design this circuit is to use a synchronous circuit with the pulse input as the clock. The pulse signal then disappears from the transition diagram since the clock is implicit. This gives the simplest circuit for this problem. The output is two variables, because the full step sequence only needs information on magnetic pole reversals to operate.

The outputs are xy:

$$A = x; \quad B = x'; \quad C = y; \quad D = y'$$

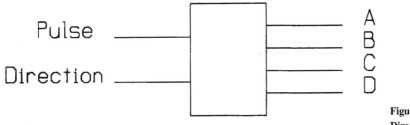

Figure 7.2 Pulse-Direction Control

7.6 VARIABLE-RELUCTANCE STEPPING MOTORS

Variable-reluctance stepping motors use a ferromagnetic rotor rather than a permanent-magnet rotor. The stator windings are excited in a sequence that will cause the rotor to move to a position that minimizes the magnetic reluctance between the stator and rotor. This is accomplished by using a different number of rotor teeth and stator poles. For any given set of stator pole windings that are excited, only a certain set of the rotor poles (teeth) will line up. When the next set of pole windings is excited, a different set of rotor poles will have to line up, causing the rotor to move one step. Because the rotor is not magnetized, it takes twice as many pole/tooth sets to get the same step size on a variable-reluctance design as it does on a permanent-magnet stepping motor.

Figure 7.3 on next page shows an idealized variable-reluctance stepping motor having six poles on the stator and four poles on the rotor. This is a three-phase motor, in which only one set of windings is energized at a time. The stator poles are labeled as to which set of windings energize that pole; with six poles, there are two poles for each winding set. As shown in part (a) of the figure, winding 1 is energized (as shown by the asterisks) and the rotor is in an equilibrium position with one pair of its poles lined up with the energized stator poles (the asterisk on the rotor is for identification purposes only; it does not imply any electrical differentiation among the rotor poles).

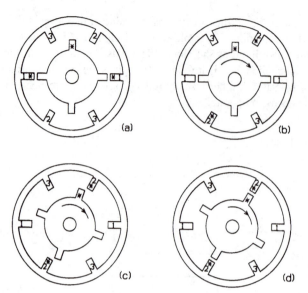

(a)   (b)   (c)   (d)

Figure 7.3 Variable-Reluctance Stepping Motor

The motor is stepped by next energizing either winding set 2 or 3, depending on the direction of motion that is desired. Winding 2 is shown energized in part (b), with the rotor not yet moved from its previous position. The rotor poles closest to the winding 2 stator poles cause a torque that will tend to rotate the rotor in the direction of the arrow. The rotor is shown on its way and at its next equilibrium position in parts (c) and (d). The total motion for the step is 30°. If winding 3 were to be energized, it can be seen from part (d) that the rotor poles closest to the winding 3 stator poles would cause another step in the same direction.

7.7 STEPPING MOTOR PERFORMANCE

7.7.1 Resonance

For step-to-step motions, the stepping motor behaves very much like a classical mass–spring–dashpot system (Figure 7.4). The spring is provided by the interaction between the rotor and the magnetic field of the stator poles that are energized. During the time period of a single step, the stator excitation does not change. When operating unloaded, the mass is the rotary inertia of the rotor. Damping comes from friction, windage, and electrical dissipation. As with any mass-spring system, it will have a resonant frequency. The frequency will depend on the load, but if excited at a stepping rate near the natural frequency, the performance will deteriorate and the motor will miss steps. Resonance is probably the major problem in stepping-motor application. It arises from the resonant structure of the rotor–stator interaction and from structural resonances excited by the discontinuous motion.

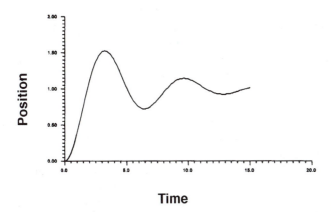

Time

Figure 7.4 Single-Step Dynamics

The two defenses against resonance are (1) avoiding stepping rates that excite resonances, and (2) increasing damping. Most stepping motors have means for avoiding specified frequencies, including switching back and forth from full stepping to half stepping (see the following sections for half-stepping control). Damping can be introduced passively with the addition of mechanical elements or electrical components. It can also be controlled actively by using velocity or acceleration feedback.

7.7.2 Static Operation

The reason that stepping motors have resonance, the spring effect introduced by the magnetic field, also provides one of their major features, a strong holding torque that will maintain position even when loaded, without feedback devices. The maximum allowable motion from application of static loads is such that the rotor stays within the range such that the next step will still provide sufficient torque to move to the next position. This is usually less than the peak theoretical torque, which occurs just before the motor slips. If slip occurs, the motor will move several steps, thereby losing the position integrity of the system. Permanent-magnet stepping motors will slip in multiples of four steps so that they end up in an equilibrium position.

Permanent-magnet stepping motors also exhibit holding torque in the absence of any excitation of the stator coils. This occurs because of the interaction between the permanent magnet and the stator. For systems that need to maintain position integrity even through power failures, if the static load is less than the *detent torque* or *residual torque*, position can be maintained with no power applied. It could be detrimental in other situations. Variable-reluctance stepping motors have no detent torque since they have no permanent magnets.

7.7.3 Dynamic Behavior

A finite stepping rate must be maintained to get the motor from place to place. At very low stepping rates, the motor behaves very much like its static characteristics, since the transient step-to-step response has a chance to die out before the next step is made. In this mode the motor can have a very low average speed but will actually jump from one position to the next and then stop to wait for the next step. Since the behavior is essentially static, the motor can be started and stopped at will, and arbitrary changes in the stepping rate can be made, as long as the maximum stepping rate is very low.

As the stepping rate is increased, the rotor does not have a chance to stop before the next step command is given. This mode is operating in a dynamic or transient mode since the excitation is changed with the rotor still in motion. Up to a limiting step rate, which is load dependent, the motor can continue to be treated as if it were behaving statically. That is, the step rate can be changed at will, even reversed, without taking any special precautions. This limiting rate is called the start–stop limit or the error-free, start–stop (EFSS) rate. The solid line in Figure 7.5 shows a typical start–stop limit line as a function of the torque the motor must supply. The maximum start–stop rate occurs when the motor is running unloaded (minimum inertia). This curve is usually given by the motor manufacturer but must be used with caution since it is affected by the nature of the load as well as its torque requirements and by the nature of the amplifiers used to drive the motor. If resonances are encountered within the operating envelope of the motor, this start–stop curve will be changed in shape, and if the resonance is only lightly damped, could be severely distorted. The stall torque and detent torque are also shown as typical points in the performance space of the motor.

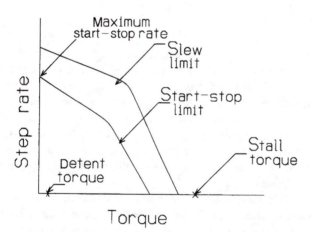

Figure 7.5 Stepping Motor Performance Limits

It is sometimes possible to operate the motor above the start–stop limit curve. If the load has a substantial inertial component, the torque required to accelerate is much higher than the torque required for cruising. Thus once the load has been accelerated, it is possible to raise the stepping rate past the start–stop limit without violating the torque limit, because less torque is needed to maintain speed. This limit is shown on the performance curve as the *slew limit*. When operating in the region between the start–stop limit and the slew limit, the step rate cannot be changed arbitrarily. Before stopping the motor, the step rate must be brought gradually into the start–stop range. If the stepping rate is changed too fast, the torque limits will be exceeded because of the inertial load, and the motor will miss steps. If the load on the motor is primarily resistive, with no significant inertial component, the start–stop range cannot be extended. The process of gradually accelerating or decelerating the motor, called ramping, is essential in high-performance stepping motor systems. The ramping rate and the maximum slew stepping rate are strong functions of the nature of the load and amplifiers.

7.8 THERMAL CHARACTERISTICS

One or more coils of a stepping motor are always energized, even when the motor is not moving. If the motor is being driven by an amplifier that acts as a voltage source, the current flowing through the coils will be maximum when the motor is not moving, because there will be no back emf to oppose the applied voltage. Therefore, the maximum heat generation also takes place with the motor stationary. The maximum heat-generation condition is also the worst case for heat removal, because with the rotor not moving, the convection cooling is at a minimum.

Temperature is thus a major operating limit for stepping motors. They are normally designed to run quite hot compared to other types of motors, but even so, temperature rise in the windings must be carefully controlled. In "normal" operating environments, observation of the manufacturer's voltage limits for the motor will usually ensure safe operation. With the thermal limit being the most restrictive operating limit, performance of a motor can be improved by providing more than normal cooling capability. The voltage, and thus the current, supply for the motor can then be increased without exceeding the temperature limit.

Another way around the performance limitations of stepping motors is to use an amplifier that controls current rather than voltage. The usual manufacturer's specifications indicate the rated voltage. This voltage is based on the temperature that would be reached in the motor when at rest for a long period of time, since stepping motors are normally used in positioning applications. However, the heating is dependent on the current rather than the voltage; thus, using the voltage rating results in a conservative design. When the motor is operating at a high stepping rate, the current at rated voltage will be less than the zero-velocity current because of the back-emf effects, and just after a change in excitation, the current will also be less than the zero-velocity current because the coil inductance will slow the buildup of current. Current amplifiers will increase the voltage in those circumstances until the rated current is reached. The torque will be substantially higher than the torque developed at the rated voltage, so that faster acceleration and higher stepping-rate limits will be possible.

7.9 SOFTWARE STEPPING MOTOR CONTROL

Stepping motor excitation can be handled completely in software if the stepping rate is low enough. A common structure is to use a clock interrupt to drive the excitation program at the stepping rate. Each time the clock interrupts, the motor is stepped by one step if it is not already at its desired position. The excitation pattern can be either four-phase or pulse/direction. The listing on the following page shows a program fragment for the interrupt service function that would handle a four-phase driver. Each time the program is called, it compares the actual position to the desired position and takes a step toward the position desired.

```
/* Interrupt Service Routine for Four-Phase Stepper */
static long dp = 0;      /*Desired position */
static long ap = 0;      /* Actual position */

/* Excitation sequence, values are the hex equivalents of the full
step sequence given above */

static int excite[4] = {0xa,0x6,0x5,l0x9}; /*
      Hex equivalent of 1010,0110,0101,1001 */
static int ix = 0;        /* Index to current excitation */

step_isr()                /* Interrupt service routine */
{
If(ap==dp)return;         /* Nothing to do */

if(ap < dp)
      {
      if((++ix) >3)ix = 0: /* Increment the index */
      ap++;                /* Increment actual position */
      }
else
      {
      if((–ix) < 0)ix = 3;  /* Decrement the index for opposite direction */
      ap–;                  /* Decrement actual position */
      }

step_out(excite[ix]); /* Send out the excitation */
return;
}
```

The resources required for this function are four bits of digital output and, ideally, a dedicated hardware clock. Communication with this function from the background, where the position and stepping rate commands originate, is through function calls to set those parameters. Positioning commands set the variable "dp," while stepping-rate changes are made by a function that changes the interrupt rate of the clock.

If a hardware clock can be dedicated to this function, it is possible on most computers to achieve stepping rates up to a few thousand steps per second without undue strain on the CPU. Since most hardware clocks have time resolutions close to 1 µs, very good resolution on stepping rate is available. Ramping the stepping rate requires more complex logic, since the current state must be preserved, but can also be accomplished within the same overall structure.

If only software clocks are available, because there are not enough hardware clocks for all the timed functions in the program, a stepping rate of up to a few thousand steps per second can still be maintained, however, the resolution of a software clock is rarely much better than 0.5 ms, so at the highest stepping rates there is only very crude control of the stepping rate. Higher stepping rates than this require some form of dedicated controller. A fast single-chip computer can probably achieve stepping rates up to 5 or 10 times faster, largely because the entire CPU can be devoted to stepping control. Rates that are yet higher require custom circuits. Commercially available stepping motor controllers provide the basic phase control from a pulse/direction input and can also provide more sophisticated control modes, including ramping and other than full-stepping modes, as well as logic to avoid specified critical speeds.

7.10 FRACTIONAL STEPPING

7.10.1 Half Stepping

Finer resolution can be obtained with stepping motors by using modified excitation sequences. The easiest to implement is half-stepping with four-phase permanent-magnet motors. In Table 7.2 the full-stepping sequence shows a pole reversal to get from each step to the next. If instead of reversing the magnetic poles, the coils are simply turned off, the motor will step to a position halfway to the next full step. This achieves a resolution of twice that available from full stepping, using the excitation sequence of Table 7.2.

TABLE 7.2 HALF-STEP SEQUENCE

A	B	C	D
1	0	1	0
0	0	1	0
0	1	1	0
0	1	0	0
0	1	0	1
0	0	0	1
1	0	0	1
1	0	0	0

AB	Right/left		CD	Top/bottom
10	NS		10	NS
00	OFF		00	OFF
01	SN		01	SN

Figure 7.6 shows the motor moving through one step using the half-step sequence. The initial position has both coils on and is at a full-step position. The excitation is then changed so that the left and right coils are turned off. The rotor will move toward the equilibrium shown, which is half of the distance to the next full-step position. A major downside to using half-step control is that the torque is not uniform from step to step. Since the half-step position uses only one coil, its holding torque is only about three-fourths of the torque at a full-step position.

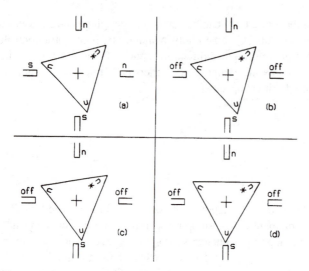

Figure 7.6 Half-Step Operation

7.10.2 Microstepping

The full-and half-step sequences can be further generalized by viewing the permanent-magnet stepping motor as a two-phase ac motor. It is two-phase because there are two sets of coordinated coils, top–bottom and right–left. The excitation for a two-phase motor is a set of 90° out-of-phase (quadrature) signals, shown in Figure 7.7. The full-step sequence is obtained by quantizing these signals to two levels, +1 and -1, shown in Figure 7.8. Only one Boolean variable is required for each phase, since there are only two possible values, -1 or +1. The sequence table lists four variables, A, B, C, and D, but each pair exists only as 01 or 10 combinations, so could be replaced with a single variable.

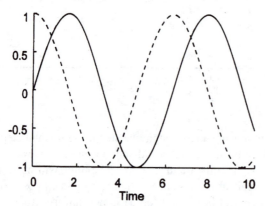

Figure 7.7 Quadrature Signals; No Discretization

One further level of quantization generates the half-step sequence, shown in Figure 7.9. In this case a trinary quantization was used. This establishes three possible values for each variable, -1, 0, or +1, so can no longer be handled by a single Boolean variable. Since the excitation table already used two variables for each phase, the half-step sequence could be done using the same table. In this case, though, the combinations 10,

01, and 00 are all used. There is no reason, however, to stop at three quantization zones. Any number of quantization zones can be used. When this is done, the stepping angle of the motor can be made arbitrarily small, yielding *microstepping* behavior. In fact, the original quadrature signals could be used as excitation sources with an analog drive system to provide continuous motion. The interpretation of the input for a stepping motor is that the horizontal axis is angle rather than time, as it is normally interpreted for an ac motor. Moving forward or backward on the angle axis will move the motor in synchrony with the excitation.

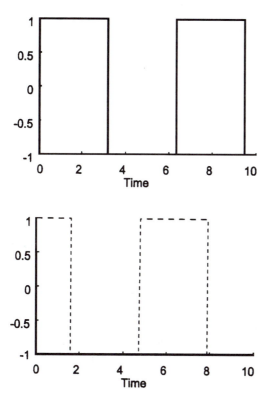

Figure 7.8 Quadrature: Binary Discretization

In microstepping operation, the relationship of stepping motors to synchronous ac motors becomes very clear. The major difference in usage is that the excitation is not a constant-frequency signal. Rather, the excitation often is constant, with the motor held in a stationary position. It is varied when the motor is to be moved to a new position. A major attraction of both full- and half-stepping is that the amplifiers that drive the coils are always either full-on or full-off. This allows the use of *switching amplifiers*, which are more efficient and less expensive than *linear amplifiers*. Linear amplifiers can produce output voltages anywhere in a specified range, whereas switching amplifiers can produce only full voltage or nothing. Implementation of microstepping requires use of linear amplifiers or of pulse-width modulation (PWM).

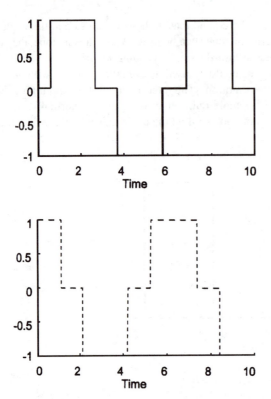

Figure 7.9 Quadrature: Trinary Discretization

Pulse-width modulation is a method that uses the natural low-pass filtering of the motor's inertia to smooth the output of a digital signal. A switching amplifier can be used, so that the output can only be either on or off. A rectangular wave is generated at a fixed frequency, with the frequency chosen to be high enough so that the motor output does not fluctuate unduly. The power to the motor is varied by changing the proportion of on-time to off-time (the *duty cycle*) of the amplifier output. The advantage of using PWM in this application is that microstepping can be achieved without using linear amplifiers. PWM can, however, generate significant amounts of audible noise and large amounts of radio-frequency interference.

7.11 TORQUE CHARACTERISTICS

The static torque characteristics of stepping motors are roughly sinusoidal with angle. The zero point in the torque curve is an equilibrium point. Static characteristics are obtained when the excitation is not being changed and when the rotor velocity is negligible. The static torque determines the holding or stall torque of the motor, as shown in Figure 7.5. The stall torque is developed by displacing the rotor away from its equilibrium. Up to a point, the restoring torque increases with displacement. If the rotor is displaced further, the restoring torque starts to fall off and the rotor will "slip" to a new equilibrium position. The rated stall torque is determined by applying a suitable safety factor to the maximum theoretical torque.

The situation gets more complicated if fractional stepping is used. In the case of full stepping, the holding torque is the same for all equilibrium positions. When fractional stepping is used, though, the holding torque differs depending on the excitation pattern at each equilibrium point. In half-stepping, the holding torque for those positions at which one set of coils is off is significantly less than the holding torque at the positions where all the coils are energized. In microstepping, the holding torque at the positions corresponding to the full- and half-step positions are the maximum and minimum holding torques developed. The holding torque at all other positions will be intermediate between those values.

When the rotor is moving, back-emf will be developed due to the motion of the magnetized rotor in the field caused by the stator coils' excitation. This back-emf appears as a voltage that "bucks" the voltage applied to the motor by its amplifier. The current through the stator coils, which to a first approximation determines the torque, depends on the *net* voltage on the motor and is thus reduced from the value it would have if the rotor were stationary. This becomes important at high stepping rates, where the rotor does not come to a stop between steps. The effects of back-emf can be minimized by using a current amplifier to drive the motor rather than a voltage amplifier, as discussed above. Current amplifiers normally operate by sensing the current and using a feedback loop to adjust the voltage so as to achieve the desired current level. This takes some time, so even with a current amplifier, there is a period of reduced torque.

Stepping motor specification requires application of the motor's torque characteristics to the load being driven. Most stepping motor manufacturers give design guidelines for driving loads that are highly inertial and for driving loads that are noninertial. More complex loads might require simulation of the system or application of differential equation models to determine optimum motor characteristics.

7.12 PROBLEMS AND DISCUSSION TOPICS

1. Equip a stepping motor (permanent magnet or variable reluctance) with a pointer or dial of some sort that can be used to measure its angular position visually. Assuming a step/direction or step-forward/step-backward interface, use a microcontroller or lab computer to send a stream of steps to the motor at a constant rate. (If only a stepping motor with a four-phase interface is available, do Problem 2, then this problem.)

   (a) Figure out the number of steps per revolution and verify that experimentally. Determine a step rate that appears to operate smoothly.
   (b) Run the step generator for many revolutions — check to see if the pointer returns to its original location.
   (c) Run the sequence forward, then backward for many repeats to see if the pointer always returns to the same place.
   (d) Increase the stepping rate and repeat parts (b) and (c). Repeat until missed steps are observed. When missed steps are first observed, is it in (b) or (c)? As the step rate is increased still further, are missed steps observed in (b) and (c) or only in one case? (If at the maximum step rate that can be achieved with software the motor still does not miss steps, switch to a function generator and see if the point at which steps are missed can be determined by the sound of the motor.)
   (e) Add an inertial load to the motor and repeat the tests.

2. Develop a logic circuit that can take step/direction inputs and produce four-phase output for a permanent-magnet stepping motor. If no motor is available, use a multichannel digital oscilloscope or logic analyzer to determine if the correct output patterns are generated.

3. Develop software that can "ramp" the step rate at a constant acceleration to a desired velocity. The ramp should start from a specified start–stop velocity and ramp up to the desired velocity. Using first a bare-shafted motor and then a motor with a substantial inertial load, find the maximum attainable velocity without missing steps. Modify the inertial load by half and then double and observe the maximum speeds as well as necessary ramp rates. Try other ramp profiles (e.g., exponential rise) and see if the performance improves or deteriorates (in terms of the time needed to reach the final velocity).

4. Design a circuit to take pulse/direction inputs and generate half-step inputs for a permanent-magnet stepper motor.

   (a) Using a static load, measure the holding torque at a full step and at a half step. Do they match expectations?

   (b) Compare the maximum start–stop rates (and associated velocities) with those achievable with full stepping (with and without a load added).

   (c) Make the same comparison for ramped operation.

5. Attach an analog tachometer to a stepping motor. Using high-speed data acquisition or a digital oscilloscope, find the actual motor velocity versus time behavior while operating at a constant step rate. Experiment with a range of step rates, including ramped rates above the start–stop limit.

6. Using a high-resolution angular position measuring instrument, find the static stiffness and associated position errors for a stopped motor. Check both full- and half-step regimes.

7. If suitable proportional amplifiers are available, develop a microcontroller or benchtop computer driver for microstepping. Again, measure the performance in start–stop and ramping modes.

8. Using a ROM as a look-up table, design a logic circuit for microstepping (including the digital-to-analog conversion for the output). Measure performance and compare to full- and half-stepping modes.

9. Arrange a torsional spring and an inertia so that a stepping motor is driving a resonant load.

   (a) Observe what happens when the step rate approaches the resonant frequency.

   (b) What happens at various ramp rates when the step rate goes through the resonant frequency?

   (c) Using a combination of full and half stepping, devise a strategy so that the motor velocity can be taken past the full-step resonance without exciting the resonance. How does the motor respond to being switched back and forth between full and half stepping?

10. Modify the resonant arrangement so that the motor is mounted on a resonant base but driving a load rigidly connected to the motor shaft. Redo the experiments of Problem 9.

8

DC Motors

Dc motors have the wonderful property of simplicity, at least as viewed by the user. When a voltage is applied to the motor, it turns. The higher the voltage, the faster it turns. This property has made dc brush motors very popular since their invention by Faraday in the mid-nineteenth century. The dominance of dc motors was challenged in the early twentieth century by the widespread adoption of ac electric power distribution. Ac motors were cheaper to build and could run off the available power supply, so displaced dc motors in large numbers of applications. The inherent controllability of dc motors, however, prevented their complete demise. Machine control, among other applications, continued to require dc motors. Developments in power electronics in the mid and late twentieth century made possible the synthesis of arbitrary, high-power waveforms. With this ability, the controllability of ac-style motors can be vastly increased. The simplest of these motors is the stepping motor, which has successfully replaced dc motors in many positioning applications. Dc-brushless motors and ac servomotors represent a convergence of this trend.

8.1 COIL–FIELD INTERACTION

As with any motor, a dc motor depends on the generation of a force when a current-carrying conductor is placed in a magnetic field. In dc brush motors, the current-carrying conductor is connected to the rotor (the armature). The stator is used to provide a magnetic field that is fixed in space. Figure 8.1 shows the simplest possible arrangement: one coil attached to the armature, free to rotate in a magnetic field supplied by the stator. In the configuration shown in (a), the rotor will tend to rotate as shown. The direction depends on the direction of current flow in the armature; if the current flow were to be reversed, the direction of the torque applied to the rotor will also change.

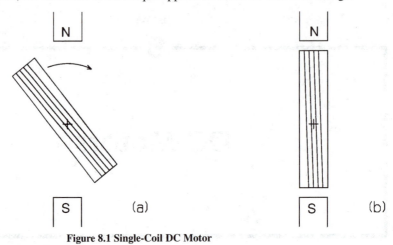

Figure 8.1 Single-Coil DC Motor

A simple motor can be based on this arrangement. If some mechanism can be provided to apply current to the armature coil only during a portion of its rotation, current can be applied only when the torque on the rotor tends to rotate it in the desired direction. The rotor will coast for the remainder of the time. This motor has several difficulties. First, if the rotor ever ends up in the position shown in Figure 8.1b, it will be impossible to apply torque to it since it is in an equilibrium position. Second, the motor will not be very effective with respect to its size, weight, or cost, because the coils are in use about only half the time. The average torque developed by the motor will be much less than it should be. In addition, if the load on the motor is primarily frictional, with very little inertial component, the rotor might not coast far enough to get into the region where the current gets turned on again.

8.2 COMMUTATION

Examining these problems one at a time, most of the features of an effective dc motor will evolve. First, a means is needed to control the current flow to the coil as a function of its position. This can be done mechanically with a commutator. A commutator uses electrical contacts sliding on the outer surface of the shaft (brushes). The shaft is split into electrically isolated regions by selective use of conducting and insulating materials. As the shaft rotates, the brushes alternately make contact with different circuits, so can change the current flow in the armature.

Figure 8.2 shows another single-coil configuration with a commutator. The commutator brushes are shown on the inside of the shaft for clarity in this cutaway drawing. In reality, they would normally ride on the outside of the shaft. The rotor shaft is broken into two electrically isolated sections by the insulating material, shown crosshatched. This configuration answers one of the questions above. It always keeps both coils activated so there is no dead time. It works by reversing the current flow through the coils as the split in the shaft goes past the brushes. The voltage source, *V*, imposes a voltage across each of the coils shown. In the configuration shown, the equilibrium point is the switching point. When the rotor moves past that point, the current will switch, thus continuing the motion in the same direction.

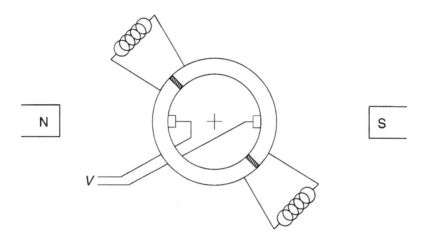

Figure 8.2 Single-Coil Motor with Commutator

The point of switch is a major problem point. The coils are inductive, so the current cannot be changed or stopped instantaneously. At the moment of switching, the brushes may actually cause a short circuit across the coil. These conditions cause some very high voltage transients, which can be observed as sparking during operation of dc motors. This electrical arc formation is a significant cause of wear in the motor, and the rapid buildup and discharge of voltages causes considerable high-frequency electrical disturbance.

This motor applies continuous torque to the rotor but still has the dead-center problem, that is, there is a point at which no torque can be applied. The torque application is also very nonuniform, since the torque on the rotor is highly dependent on the rotor's position. The solution to this is to increase the number of coils and the number of splits in the shaft. Figure 8.3 shows a four-coil version of the same motor. In the configuration shown, the two upper coils are connected in one direction across the voltage source, and the two lower coils are connected in the other direction. The circuit starts from one of the brushes, continues into the conducting portion of the shaft, and goes through a coil into the conducting section of the shaft on the other side of the coil, through the next coil, then back to the other brush. Every time one of the insulating gaps crosses the brushes, one set of coils reverses the sense of its magnetization. With the brushes oriented as shown, whenever a set of coils reaches the dead-center position, their polarity is reversed. When that happens, however, the other two coils are at 90° to the dead-center position and continue to supply torque to drive the rotor. As the number of coils is increased, the smoothness of the torque application also increases. In the limit, the armature would maintain a constant magnetic angle to the applied field, so the torque would not vary with rotor position.

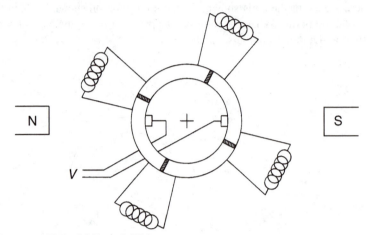

Figure 8.3 Four-Coil Motor

8.3 OPERATING CHARACTERISTICS

The most fundamental operating characteristics of a dc motor can be derived from a "black-box" view of the motor. In its simplest incarnation, a motor is a box with two wires coming out of one side and a shaft coming out of the other side (Figure 8.4). Other than the electricity through the wires and the mechanical motion of the shaft, no energy or material crosses the boundary of the motor's boundaries. For this conceptual analysis, thermal effects, airflow, bearing friction, and so on, are all being neglected. Furthermore, it will be assumed that there is no mechanism for storage or dissipation of energy inside the box labeled *Motor*. These assumptions may seem very simplisitic, but they will produce the basic structure and most important operating parameters of dc motors. Because of this, the power flow on the electrical side must instantaneously equal the power flow on the mechanical side:

Figure 8.4 Black-Box View of a Motor

Mechanical power = electrical power

Power flow on each side is the product of the variables listed:

$$P = Vi = \tau\Omega \qquad\qquad (8.1)$$

Electromagnetic theory relates the torque to the current,

$$\tau = K_\tau i \qquad\qquad (8.2)$$

for the case where the magnetic field imposed by the stator does not change. Substitution into the power equation then indicates that the voltage must be proportional to the speed,

$$P = Vi = \tau \, \Omega = K_\tau i \Omega \qquad (8.3)$$

so

$$V = K_\tau \Omega \qquad (8.4)$$

This equation is the origin of the phenomenon of back-emf, the voltage produced by the motor as a result of its speed. This is usually written as

$$V_b = K_E \Omega \qquad (8.5)$$

The torque constant and the voltage constant are really the same. They differ only if inconsistent units are used for the power variables, which is usually the case.

8.4 MOTOR-DRIVEN SYSTEM

With these definitions, operating equations can be derived for a typical, simple motor system (Figure 8.5). The motor is driven by a voltage, with a series resistance representing the motor winding resistance and resistance of the external circuit elements. The inductance of the winding is ignored (when considered, it appears in series with the resistance). The load is a flywheel, that is, a pure inertial element. The mechanical side is governed by the rotary equivalent of Newton's law, relating the torque to the acceleration,

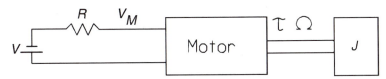

Figure 8.5 Motor with Drive and Load

$$\frac{d\Omega}{dt} = \frac{\tau}{J} \qquad (8.7)$$

The torque is related to the current by equation (8.2), but the current depends on the voltage drop across the resistor,

$$i = \frac{V - V_m}{R} \qquad (8.7)$$

and the voltage across the motor is related to the speed by equation (8.5), so the current becomes

$$i = \frac{V - K_E \Omega}{R} \qquad (8.8)$$

Substituting in equation (8.6) for the torque in terms of the current gives

$$\frac{d\Omega}{dt} = \frac{K_\tau (V - K_E \Omega)}{R} \qquad (8.9)$$

This equation has an equilibrium point, the speed at which the voltage across the motor becomes zero. This happens when the back-emf is equal to the voltage applied:

$$\Omega_{eq} = \frac{V}{K_E}$$

(8.10)

When this value is substituted into equation (8.9), the right-hand side becomes zero, verifying the equilibrium. Thus for any given input voltage, the motor will seek a constant operating speed, independent of the inertia of the load.

The load inertia will affect the transient behavior. If power is applied to a stationary motor, its speed will asymptotically approach the equilibrium velocity. Figure 8.6 shows the rise of the motor velocity with time when a motor is started from rest. The two curves represent two values of inertia, one twice the other. Note that the response time changes, with the low-inertia case having the faster rise time, but the equilibrium value does not.

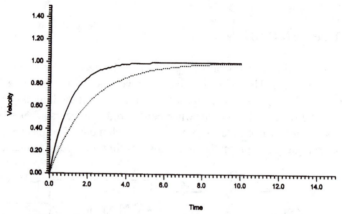

Figure 8.6 Velocity Response of Motor: Voltage Input

8.5 GENERATORS/TACHOMETERS

The power-flow-based equations derived above have no bias with respect to a preferred direction for power flow. In a motor, the usual power flow direction is from the electrical side to the mechanical side. The same mechanism, however, can operate the other way also, with a mechanical drive to produce electrical power out. A device operated this way is a generator and can be used to drive an electrical load. When operated to provide braking, dc control motors are really acting as generators, although efficient electrical energy production is not a major design concern.

An important application of dc motors that are run as generators, is their use as a velocity-measuring instrument, a *tachometer*. The equations above describing motors also apply to generators and tachometers. Tachometers, however, have special needs. First, no significant electrical power is drawn from a tachometer. It is normally run with the electrical circuit connected to a high-impedance device, so is, essentially, run open circuited. Thermal considerations of the sort needed to protect a motor from burning out are not relevant. On the other hand, as a measuring instrument, the voltage constant must be very stable with temperature so that the measurement of the velocity does not drift. The output voltage should also be free of ripple, so a high number of armature coils is called for. To reduce noise in the measurement further, the brushes must be constructed to minimize noise due to the sliding contact and polarity reversals.

8.6 BRUSH PROBLEMS

Despite their simplicity, the brushes that provide both the "computation" of where the switch should occur, based on the commutation point locations, and the power switching have problems. They tend to generate large amounts of electrical noise due to the rapid switching and the arcing that takes place through a switch, and they wear out, increasing the maintenance requirements and operating costs. They are also only capable of applying on–off voltages. The physical characteristics of the brushes tend to be the limiting voltage factors as well. A response to this has been the development of brushless motors, which are discussed later in this chapter.

8.7 DC MOTOR TYPES

8.7.1 Field–Wound Motors

The first distinguishing feature of dc motors is the way in which the magnetic field is generated. A field-wound motor uses another coil on the stator as an electromagnet to produce the field. If the same voltage supply is used to excite both the armature and the field, the motor is referred to as a self-excited motor. Self-excited dc motors include series wound, with the field and armature windings in series, shunt wound, with the field and armature windings in parallel, and compound motors, which include both series and shunt field windings. Reversing a self-excited motor adds an element of complexity because the armature (or field) must have its polarity reversed separately.

The various winding patterns can be used to affect the shape of the torque–speed curve. Series-wound motors tend to have a high starting torque that drops off rapidly, shunt-wound motors have a flatter initial characteristic and then a rapid drop-off, while compound motors can be customized to a combination of the two. Motors with constant field excitation have approximately linear torque–speed curves. This can be derived by combining equations (8.2) and (8.8), which were derived on the basis of constant stator field. Field-wound motors have the disadvantage (compared to permanent-magnet motors) that there is substantial heat generation associated with the field windings.

8.7.2 Permanent-Magnet Motors

A common way of getting a constant stator field is to use a permanent magnet as part of the stator, to get a permanent-magnet (PM) motor. Permanent-magnet motors have become the first choice for many control applications. In high-current high-torque situations, the ability to maintain the stator field without using any electrical power reduces the thermal load significantly, thereby improving motor output and reliability. Permanent-magnet motors can also achieve similar output torque with much smaller frame sizes because of the high magnetic field that can be produced with ceramic magnets. These, combined with the linear torque-speed relationship, make permanent-magnet motors very attractive.

Permanent-magnet motors can be demagnetized by excessive current flow in the armature (maybe they should be called semipermanent). The demagnetization effect is enhanced at high temperatures, which are also associated with high load conditions. In normal operation, the armature current acts as an electromagnet opposing the magnet field of the permanent magnet in the stator. This is usually a reversible effect; when the armature current is removed, the magnet recovers its full strength. If the current exceeds a limiting value, however, there can be permanent loss of magnetization, resulting in deteriorated performance of the motor.

To construct a dc motor with maximum torque, the magnetic flux should be maximized, since the torque is proportional to magnetic flux. Magnetic flux depends on the arrangement of the magnetic elements and the materials through which the magnetic field passes. Most motors are made with steel in the magnetic circuit because of its very high permeability and thus high resulting magnetic flux. This leads to relatively high inertia values for the armature because of all of the steel in it; the design is very effective, however, for driving loads that are also high inertia since the overall system performance is maximized. The use of gears, lead screws, belts, and so on, can match the reflected inertia of the load to the inertia of the motor to further optimize performance.

8.7.3 Ironless-Core Motors

In some applications, though, the load is very low inertia and maximum acceleration is the primary system performance index. In these cases, the maximization of flux, and thus torque, is not as important. A motor built with no steel in the armature, an ironless-core motor, can be optimized for maximum acceleration of very low inertia loads. These motors are usually built in either a pancake arrangement, with a flat armature, or with a shell armature. In either case, no iron or steel is used in the construction, and the armature is kept as light as possible.

8.7.4 Linear Motors

Standard motors produce rotary motion. Linear motion is often needed for applications, and can be created from rotary motion with rack and pinion, lead screws, belt drives, chains, and so on. These mechanisms, each has its own peculiarities, introduce side effects into the system. It is possible to produce linear forces directly from dc motors, thereby avoiding the side effects of the rotary-to-linear converters. A linear dc motor can be visualized by "unwrapping" the motor of Figure 8.3. The moving carriage, or head, the linear equivalent of the rotor, moves over the full length of the motor (Figure 8.7). In the design shown in the figure, the channel is assumed to be fixed, and the permanent magnets are attached to the moving head. The coils are wired to electrically isolated sections of the fixed channel. As the head moves along, brushes on the head activate the coils closest to the head. The figure shows a conceptual schematic view, so the geometric relationship of the coils and head is not intended to be accurate. Since the head is moving, the power for the coils must be brought onto it for distribution to the brushes. This is done in the design illustrated with a second set of brushes, the power brushes, that are in contact with a power bus along one side of the channel. For relatively short motions, wires can be used to power the coils. Other linear motor designs are possible, with the magnets on the fixed channel and coils on the head, for example. If the motor is configured with the head fixed and the channel moving, the power for the brushes could be directly attached to the head and no power brushes would be needed.

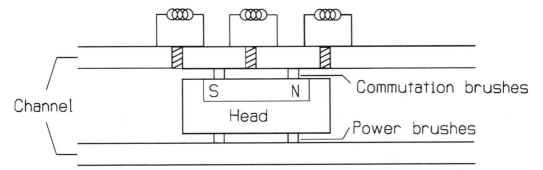

Figure 8.7 Schematic View of Linear Motor

Linear motors have the tremendous advantage of being able to apply a force directly to the object being moved. The suspension of the load can be done without having to worry about transmitting torques or forces from the motor. For very low friction, for example, air bearings can be used to suspend the load. The only frictional connection of the load to its base will be the motor's brushes. There are some disadvantages, however. Unlike a rotary motor, the size of the motor depends on the distance to be moved, so, for linear motors with long paths, the motor can be very expensive. With a rotary motor, only the transmission mechanism has to extend the length of the linear travel. Also, the forces developed are only as a result of the interaction of the coil on the moving carriage with the local stationary magnetic field producing elements, whereas the torque production in a rotary motor is always assisted by all its elements. The result is that the forces generated are less than could be generated by an equivalent rotary motor, and there is a strong potential for local overheating. Finally, since the linear motor acts directly on the load, there is no convenient way to use transforming elements (gears, for example) to match the motor's impedance to the load's (impedance in this context refers to the force–speed relationships).

8.8   IMPEDANCE MATCHING

The torque–speed characteristics of motors are fixed by their design parameters. In general, there are many more potential load situations than there are standard motors available, so it may not always be possible to match a motor optimally to its intended load. A gearbox or other impedance-transforming device, such as pulleys or lead screws, can be used to improve the match. These devices are the mechanical equivalents of electrical transformers. The general arrangement is shown in Figure 8.8, where the "gearbox" could be any transforming element. The gearbox or other transformer is first used to match the needed static characteristics of the motor–load system. The speed range and torque range are the most important of these static properties. Within those constraints, however, many motor–gearbox combinations are possible. Is it possible to optimize further? If maximum power transfer to the load so as to achieve maximum acceleration is the most important dynamic operating characteristic, the classic impedance-matching rule can provide further optimization. For this case it will be assumed that the load is a pure intertia and that the motor itself is also a pure inertia. Losses due to friction, compliance, and so on, will be neglected.

Figure 8.8 Gearbox for Impedance Matching

8.8.1 Optimum Gear Ratio

The total inertia as "seen" by motor (i.e., reflected through the gearbox) is

$$J_T = J_m + \frac{1}{r^2}J_L \tag{8.11}$$

where **r** is the gear ratio. The kinetic energy of the load is

$$KE_L = \frac{1}{2}J_L\Omega_L^{\ 2} \tag{8.12}$$

which can be differentiated with respect to time to get power:

$$P = \frac{d}{dt}KE = J_L\Omega_L\frac{d\Omega_L}{dt} \tag{8.13}$$

Reflect this back to the motor:

$$P_L = \frac{J_L\Omega_m}{r^2}\frac{d\Omega_m}{dt} \tag{8.14}$$

The current during this process can be considered to be constant, since maximum power can be achieved only by delivering maximum current to the motor. The (constant) acceleration is proportional to applied torque divided by total inertia:

$$\frac{d\Omega_m}{dt} = \alpha m = \frac{\tau_m}{J_T} \quad \Omega_m = \alpha_m t \tag{8.15}$$

Since the acceleration is constant, the power to the load is

$$P_L = \frac{J_L\tau_m^2 t}{r^2 J_T^2} \tag{8.16}$$

Optimize this by finding a maximum P_L value with respect to the gearbox ratio, r.
Only the denominator depends on r, so differentiate with respect to r and set to zero:

$$\frac{\partial\left\{r[J_m+\frac{1}{r^2}J_L]\right\}^2}{\partial r} = 0 \tag{8.17}$$

This is zero at

$$Jm - \frac{1}{r^2}J_L = 0 \quad J_m = \frac{1}{r^2}J_L \tag{8.18}$$

But this is just the load inertia as reflected at the motor. In other words, adjust r so that when viewed from the motor, the impedance (inertia) of the motor matches the impedance of the load as viewed through the gearbox.

8.9 THERMAL CHARACTERISTICS

The main enemy of motors is heat. High temperatures can destroy the components of the motor, and even if they do not cause permanent damage, they can change the operating parameters and cause significant deviations in motor performance. The heat results from mechanical and electrical dissipative mechanisms and appear in the motor's dynamic equations as a force (or torque) that resists the motion (i.e., a damping force). The main sources of heat production are winding electrical losses (i.e., resistive heating), eddy currents, magnetic hysteresis, windage, friction, short-circuit currents, and brush contact resistance. The dissipation in the windings comes from the resistance of the coils and the associated ohmic heating. A field-wound motor will also have ohmic loss from the field coils. They are proportional to the square of the current flow and rise rapidly as the motor is loaded.

The eddy current losses are associated with currents induced in the iron core due to the changing magnetic fields, while the hysteresis losses, also associated with the iron core, come from the shifting of magnetic domain boundaries as the armature rotates. Both of these are motor-speed dependent rather than load dependent. Ironless-core motors do not exhibit either phenomenon, so can be made more efficient than iron-core motors.

Windage and friction cover mechanical sources of dissipation and also tend to be speed dependent. They are heavily dependent on the specific motor design. The brushes contribute to two loss mechanisms. They contribute mechanical friction, and there is also electrical dissipation due to the sliding interaction between the brushes and the shaft. The brush contact resistance is a complex combination of factors, including temperature, speed, and armature current.

Short-circuit currents occur every time the commutator moves from one sector to the next. They are a combination of large currents caused by the rapid change in voltage on the coils, which act as inductances, and the momentary short circuit that occurs when both ends of a coil are connected briefly to the same side of the electrical circuit. The amount of loss is most closely correlated with motor speed. The operating temperature inside a motor depends on the amount of heat generated by these dissipative mechanisms and the heat transfer properties of the motor. For heat transfer purposes, the motor is usually divided into two "lumps," the armature (rotor) and the housing, because the heat transfer between them and between each of these elements and the ambient are the critical heat transfer paths. If linear heat transfer relations are used, the equations describing the heat transfer are (subscript r refers to the rotor, h to the housing, and a to the ambient)

$$C_r \frac{d\Theta}{dt} = D_r - Q_{rh} - Q_{ra} \tag{8.19}$$

$$C_h \frac{d\Theta}{dt} = D_h + Q_{rh} - Q_{ha} \tag{8.20}$$

The terms represent dissipation in the rotor and housing. For permanent-magnet motors, there is little or no dissipation in the housing because no electric current is needed to maintain the magnetic field. The simplest representation of the heat transfer is that it is proportional to temperature differences, so the heat transfer terms become

$$Q_{rh} = K_{rh}(\Theta_r - \Theta_h) \tag{8.21}$$

$$Q_{ra} = K_{ra}(\Theta_r - \Theta_a) \tag{8.22}$$

$$Q_{ha} = K_{ha}(\Theta_h - \Theta_a) \tag{8.23}$$

The use of transient heat transfer equations is very important for control applications, because there may be modes of operation in which high torques, and therefore currents, are needed only for short times. In particular, when these times occur at low speed, when the speed-dependent losses are low, it is often possible to achieve torque and current performance much greater than the "rated" values without damaging the motor.

8.10 MOTOR CONTROL

8.10.1 Voltage Control

The easiest way to control a dc motor is to control the voltage applied to it. The basic response for a voltage input was given previously. When only rough control is needed, this will do fine. However, the final speed will be dependent on the load and, to some extent, operating temperature, since the temperature will affect the motor's operating parameters. The applied torque will change with speed, and there is no way to control position at all.

Applications involving physical stops, or limit switch position detection, for example, can make effective use of this mode if the precision requirements are not too great. With a limit switch, for example, the rough speed of the motor can be controlled with the voltage, and when the switch trips, the motor power is turned off. If the load is largely frictional, the motor will stop rather quickly when the power is shut off. If the load is largely inertial, a brake may be necessary to stop it and hold it in place.

8.10.2 Torque–Current Control

Torque is a difficult variable to control directly because it is hard to measure. Either the shaft must be instrumented with strain gages to measure its angular strain, from which the torque can be computed, or the motor system must be "floated," using bearings, so all of its reaction torque to ground goes through a single member. That member can then be instrumented to measure torque. Neither of these methods is very convenient, so torque is rarely measured in working systems. As an alternative, equation (8.2) indicates the close relationship between torque and motor armature current. If the torque constant of the motor is known, this formula gives a good approximation to the torque applied to the mechanical system, including the bearing friction, brush friction, windage, and so on. It thus includes several factors that are internal to the motor, as well as the torque applied to the external load.

Current control is usually achieved by using a current-controlled amplifier. In these amplifiers, a feedback loop is implemented that utilizes a measurement of the current

(Figure 8.9). The input to the amplifier, usually a voltage, is interpreted as proportional to the desired current. The voltage applied to the motor is then manipulated so as to keep the current near its desired value. The feedback loop inside a current-controlled amplifier serves two purposes. The first is to provide static compensation to keep the current at the desired value during steady-state operation. This compensates primarily for the motor's back-emf, raising the voltage applied to the motor high enough to overcome the back-emf. The dynamic performance is also improved. The inductance of the armature coils acts to impede the initial flow of current. The feedback loop in the amplifier will temporarily raise the voltage to speed up the current's rate of change.

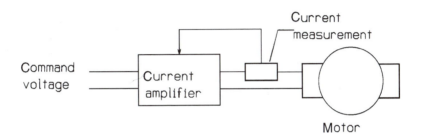

Figure 8.9 Current-Controlled Amplifier

8.10.3 Velocity Control

Control of the motor's speed or position is also done with feedback. A variety of configurations are possible for velocity control. The feedback measurement can be accomplished, for example, by using a tachometer (as described previously), by computing velocity from position measurements, or by using a resolver. For the moment it will be assumed that a "perfect" instrument is available to measure the velocity. Either voltage or current amplifiers can be used to convert the controller command output to a power input to the motor. Because of the back-emf characteristics of the motor, there is a significant difference in the resulting performance.

The basic block diagram for a speed control system is shown in Figure 8.10. The dynamics of the system are described by equation (8.6) and its related equations, as long as it is assumed that the load is purely inertial and can be considered part of the motor inertia, J. As a first approximation, all other dynamics can be ignored (motor inductance, resonances, etc.). The performance of this system depends on the controller *gain* and the type of amplifier used (Figures 8.11 and 8.12, simulated behavior). In Figure 8.11, which utilizes a voltage amplifier, the speed never quite reaches the desired voltage. In the second, where a current amplifier is used, the response converges exponentially to the desired value. The higher the controller gain, the faster the response and the smaller the final error (for the voltage amplifier case). For the simple representation of the motor system used in this example, that's all there is to it. No gain is too high!

In real systems, however, other dynamics are present also. For low values of controller gain, these other dynamics will not contribute significantly to the overall behavior. As the gain is increased, however, their contribution becomes more important, until at some value of the gain, the performance is dominated by the unmodeled (or parasitic) dynamics and a gain limit

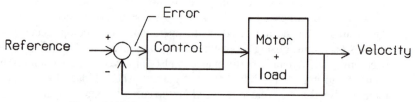

Figure 8.10 Velocity Control

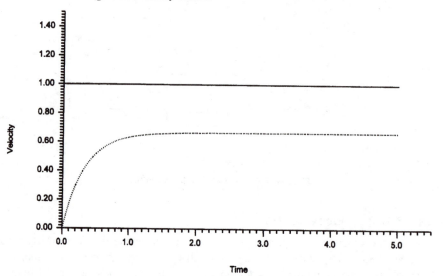

Figure 8.11 Velocity Control Simulation: Voltage Amplifier

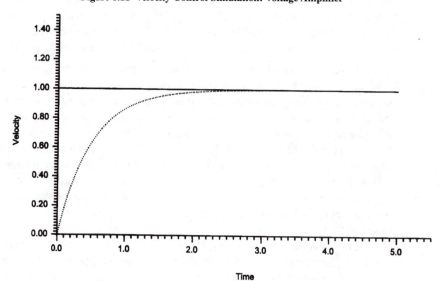

Figure 8.12 Velocity Control Simulation: Current Amplifier

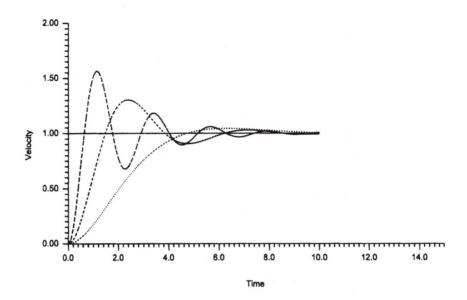

Figure 8.13 Gain Limit with Parasitic Dynamics

is reached. Figure 8.13 on page 127 shows the simulated behavior of a system that includes dynamics representing the armature coil inductance. The curves are for increasing values of the controller gain; all cases use a current amplifier. As the gain is increased, the response becomes more violent, with a large overshoot.

8.10.4 Dynamic Compensation

As the dynamics get more complicated, with significant resonance frequencies near the operating frequency, for example, the design of the control algorithm (the mathematical function that goes where the "control" element is in the block diagram) becomes more sophisticated and involves use of more control system theory. In this case, for example, the effects of the coil inductance can be offset by the use of a lead–lag dynamic compensator in the control algorithm. A dynamic compensator makes use of past and present values of the measured variable (velocity, in this case) to try to keep the control from under- or overreacting. Design of such compensators is beyond the scope of this discussion (see any control system text) but is introduced here to show the relevance of such techniques to basic velocity control. Figure 8.14 on the next page shows the simulated response when a lead–lag element is added to the controller for the highest-gain (most overshoot) case of Figure 8.13. The response is now much better behaved.

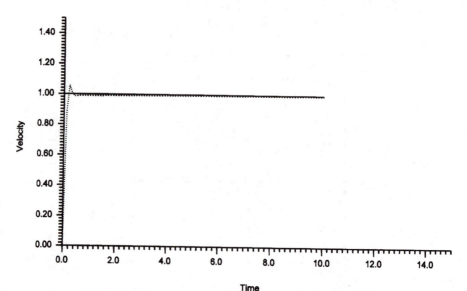

Time

Figure 8.14 Lead–Lag Compensation for Coil Dynamics

8.10.5 Position Control

Controlling position with a dc motor requires a position-measuring instrument, so a feedback loop can be implemented. The best control quality is obtained with a cascade control system, consisting of an inner velocity loop and an outer position loop (Figure 8.15). The inner (velocity) loop is the same control discussed previously. In this case the setpoint (reference value) for the velocity loop comes from the outer, position loop. The cascade control, however, uses two instruments, a velocity-measuring instrument and a position-measuring instrument. It is possible to implement a position control system with just a position-measuring instrument, as in Figure 8.16. Depending on the system performance demands, the control algorithm in systems using only position-measuring instruments can make up for the missing velocity by using additional computational elements, such as the lead–lag filter mentioned above, or a state estimator to get an approximation to the velocity.

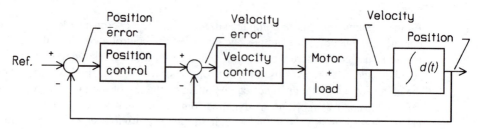

Figure 8.15 Cascade Position Control

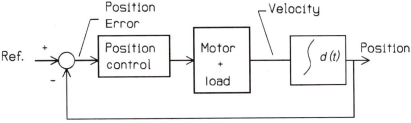

Figure 8.16 Position Control Loop

Figure 8.17 shows the simulated response of a position control system using full cascade control. As before, it is assumed that "perfect" instruments are used and that a current amplifier with no dynamics is used for actuation. The reference signal is a step. The position is thus intended to go from one stationary point to another. Called *step and settle* control, this is used in applications where a mechanism must be moved from one position to the next and then halted for some task to be accomplished. The response is fast and settles to its new position quickly. In contrast, the first attempt at using only a position measurement (Figure 8.18 on the next page) is a disaster! In this simulation, a current amplifier is used and a gain is implemented in the position loop. The response is in the form of an undamped harmonic oscillator. Changing the controller gains does not help. The only property of the response that changes is the frequency of the oscillation.

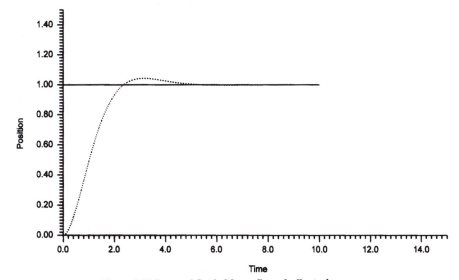

Figure 8.17 Step and Settle Move: Cascade Control

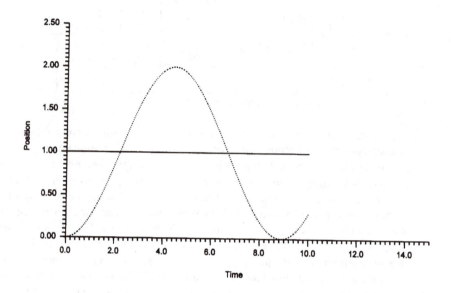

Figure 8.18 Step and Settle: Position Loop Only

A real system would not be quite as bad, since there is always some dissipation that has not been accounted for in this simulation. Friction from bearings and brushes, electrical dissipation, eddy currents, and so on, all contribute some damping to the system. Nonetheless, the performance would be unsatisfactory since the stabilization would be due entirely to elements beyond the influence of the controller, so could not be changed by changes in controller gain.

Examination of the mathematical structure of the position loop control shows that insertion of the current amplifier removes the natural damping effect of the back-emf. The back-emf is a naturally stabilizing influence because it tends to resist changes in speed: as the speed increases, so does the back-emf, which reduces the current and thus the torque. Figure 8.19 shows the simulated response of step and settle control with a voltage amplifier in place of the current amplifier. The response in this case has satisfactory characteristics but is relatively slow. If high-speed performance is not necessary, this simple configuration could be quite satisfactory. Again, control theory provides tools to improve the performance. Figure 8.20 shows the response when the current amplifier is used, but a lead–lag element is included in the controller as a compensator. The response is now fast and has only a small overshoot. Although the final recovery to the setpoint is fairly slow, the error is very small.

As has been seen in both the velocity and position control, a variety of configurations are possible. In general, the best performance is obtained by using as many instruments as possible and using the most direct actuation possible. It is usually possible

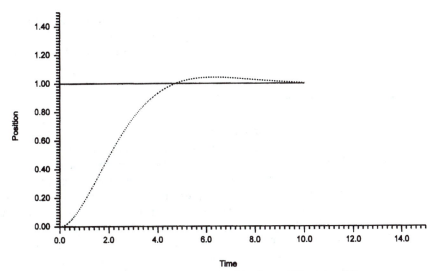

Figure 8.19 Step and Settle: Position Loop, Voltage Amplifier

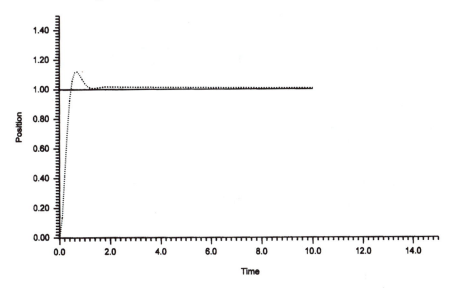

Figure 8.20 Step and Settle: Position Loop, Lead–Lag Compensation

to improve performance by using more complex control algorithms, but there is an expense. As the complexity is increased, the control system tends to become more sensitive to changes and errors in the parameters of the system being controlled. Unmodeled dynamics or changing values of friction, for example, could cause significant changes in overall system performance.

These simulations also assumed the use of perfect instruments, actuators, and computational elements. In the real world, deviations from perfection cause deterioration in performance. Noise, delays, sampling, quantization, and so on, are all present to some degree or another. Some of these effects are discussed in later chapters. Sampling and quantization have already been discussed with regard to the properties of control computers.

8.11 BRUSHLESS MOTORS

Brushes have long been recognized as the weak point in dc motors. The quest for a means of eliminating brushes has been a long one. The development of effective low-cost solid-state electronic power switches has been the liberating element. The idea is to build a motor that has the same general performance specifications as a brush dc motor but has no brushes. Figure 8.21 shows the general shape of torque–speed characteristics for dc motors, stepping motors, and ac induction motors. The dc motor characteristic is by far the easiest to deal with in a control situation that requires operation over the full torque and speed range of the motor. By contrast, both of the other motors introduce complexities into the control due to the irregularities in their torque-speed characterisitics. A brushless dc motor should look as much as possible like the torque–speed characteristic of a brush-type dc motor.

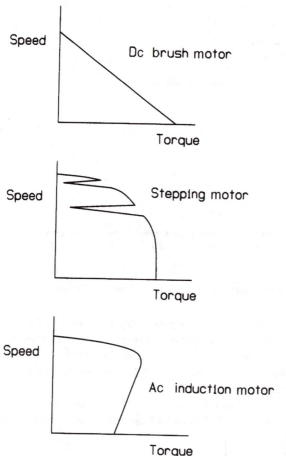

Figure 8.21 Torque–Speed
Characteristics for Various Motors

8.11.1 Brushless Motor Configuration

Figure 8.22 shows a schematic of a permanent-magnet brushless motor configuration. It reverses the configuration shown for the brush-type dc motor by putting the permanent magnet on the rotor instead of the stator. The stator windings are excited in turn as the rotor moves through successive angular positions. To implement this brushless motor, the measurement, computation, and actuation have been separated into unique components. In the brush motor, the unit consisting of brushes and commutator ring provides all three of those functions. For the brushless motor, there must be an explicit instrument to measure the position of the rotor, a computational element to figure out which windings should be excited and a power-switching device to carry out the selective actuation. The excitation uses three-phase dc signals, as shown in Figure 8.23. The excitation is a function of rotor position and requires three separate channels of power amplification (switching). This form of on–off excitation just needs discrete points of measurement, for which Hall effect sensors (magnetic flux detectors) have been commonly used.

8.11.2 Pros and Cons of Brushless Motors

No brushes. What more is there to say! Less maintenance, less electrical noise, and higher voltages can be used. On the other hand, there are some drawbacks. The control and power circuitry is much more complicated, and a position measurement is required.

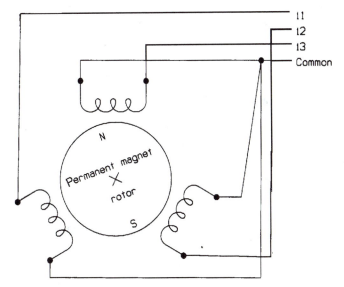

Figure 8.22 Brushless Motor Configuration

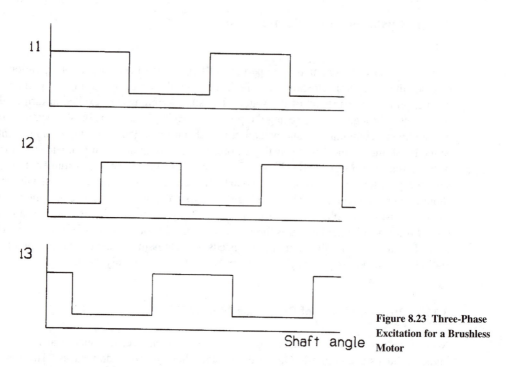

Shaft angle

Figure 8.23 Three-Phase Excitation for a Brushless Motor

The brushless motor will normally have a higher torque ripple than a brushed motor. This is because it is not practical to use more commutation points. Each additional commutation point requires an additional full set of electronics. Torque ripple in brushless motors can be reduced somewhat, however, by using a sinusoidal excitation instead of on-off excitation. This requires still more complex electronics—a linear or PWM amplifier instead of the simple switching amplifier— but can achieve better performance.

8.12 COMPUTER INTERFACE REQUIREMENTS

8.12.1 Supervisory Control

The simplest form of computer interface is for supervisory control. In this mode the actual feedback control is handled by analog computing elements and the computer is used to supply the set points to the analog controllers. The demands on the computer are for computation of the set points and communication of that information to the controller. How difficult that is depends entirely on the nature of the task. A multiaxis robotic manipulator, for example, will put a large computational load on the supervisory computer, whereas a simple Cartesian machine tool might not require much computation at all. Communication with an analog control system is most easily accomplished with digital-to-analog converters, with the analog signal representing the set point. Other communication methods are possible but would require specialized equipment as part of the analog control system as well as in the supervisory computer.

8.12.2 Direct Digital Control

Analog computing elements for the control loop have the advantage of high bandwidth and no quantization, but analog components must be hardwired, the variety of computational functions available is limited, and noise, drift, and environmental sensitivity make repeatable control difficult. Direct digital control replaces the analog elements in one or more of the control loops and provides the advantage of the computational complexity that is easily implemented digitally, and the algorithms can readily be changed. The primary performance limitation is the loop computational speed of the control computer. The control performance is strongly affected by the sampling time, which is mainly limited by how fast the computation can be completed. There is also a delay associated with the computation. If the instrument readings are sampled at the beginning of the computation period, the actuation output will not be ready until the computation is done. This delay also causes performance deterioration.

Loop sample times necessary for satisfactory performance can range down to well under 1 ms for some systems and as low as 100 μs or less for such fast systems as disk drive head-positioning control. At these sample rates, conventional microprocessors can do only simple control algorithms, which to some extent nullifies one of their major advantages. Special-purpose processors such as digital signal processors, can be used to great advantage in high-performance position and velocity and position control systems. For higher-valued applications, processors with integrated floating-point arithmetic support can be used to great advantage. When the loop control is done digitally, the supervisory control can be done in the same computer if computation time is available, or in a separate computer. If separate computers are used, a communication channel is necessary between the two.

8.12.3 Instrument and Actuation Interfaces

The complexity of the interface between the control computer and its instruments and actuators depends on whether the external device can simply be sampled whenever the computer is ready to get a value, or whether it requires servicing regardless of the state of the control program. A dc tachometer, for example, has an output voltage proportional to its speed. An analog-to-digital converter can be used to sample that voltage on demand. The computational load is small and does not require any coordination beyond whatever scheduling is used for the control loop tasks. An incremental encoder, on the other hand, or any other form of pulse-generating device must be serviced when its next transition takes place, which is normally asynchronous with the control computations, and also, normally much faster. This complicates the software design because of the multiple-event coordination.

Actuation interfaces have the same problems. A linear amplifier, for example, requires only an analog signal to drive it, which can be generated with a digital-to-analog converter. If a switching amplifier is used, it would normally be driven with a pulse-width-modulated signal. Generation of a PWM signal directly from the computer again requires timing and coordination based on the frequency and duty cycle of the signal.

Servicing high-speed devices such as encoders and PWM signals puts a heavy load on the control computer in terms of its ability to answer the interrupts that are used to schedule these events. There is a certain amount of overhead associated with answering interrupts, so that very frequent interrupts can cause a substantial loss in computing efficiency. If this is a problem, it is common to assign those tasks to other processors, some to other computers, some special-purpose devices. A communication channel must then be established to the control computer so that the device can be queried or commanded. Since querying or commanding external devices can be done synchronously with the control, the software complexity is reduced. The communication channel, however, also has software and timing requirements which must be considered in assessing the overall system performance that can be obtained when external parallel processing is used.

A further important consideration when designing the actuation–instrumentation– computer interface is the distance between the computer and the devices it must control. Analog signals, for example, are easily contaminated with electrical noise and need separate wires for each signal. Analog signals are thus less than ideal for long-distance connections. Various digital communication media can be used for longer distances or to multiplex several signals onto a single channel, but they usually require more complex equipment at both the receiving and transmitting ends.

8.13 PROBLEMS AND DISCUSSION TOPICS

1. It was shown in equation (8.18) that maximum power transfer to the load was achieved by matching the load and motor impedances. Apply this general concept to the electrical system shown in Figure 8.24 and prove that $R_L = R_s$ maximizes the power delivered to the load, R_L.

2. The motor–drive–load system was described by equation (8.9), which is repeated here:

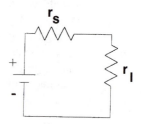

Figure 8.24 Maximum Power

$$\frac{d\Omega}{dt} = \frac{K_\tau(V - K_E\Omega)}{RJ} \qquad (8.24)$$

(**a**) Let a step input voltage be applied to a system of this sort which is initially at rest, that is,

$$\Omega(t = 0) = 0 \quad V(t) = 0 \ \text{for} \ t < 0 \quad V(t) = V_0 \ \text{for} \ t \ge 0$$

Verify that the response of the system is given by (it is assumed that the torque and back-emf constants are given in consistent units)

$$\Omega(t) = \frac{V_0(1 - e^{\frac{-t}{\tau_c}})}{K_\tau} \qquad \tau_c = \frac{JR}{K_\tau^2} \qquad (8.25)$$

(**b**) The quantity τ_c, the time constant, is a measure of the speed of response of the system. Determine the amount of time in terms of time constants that it takes for the system to reach 90% of its steady-state value. What is the effect on that time of doubling the inertia, J? Of doubling the resistance, R? Of doubling the torque constant, K_τ?

3. It is sometimes necessary or desirable to estimate the coefficients of the motor model equation from experimental data. This may be because the manufacturer's data are unavailable or the actual motor parameters differ from the published data (from aging, unknown loads, etc.). To do this it is convenient to rewrite the equation used in Problem 2 as

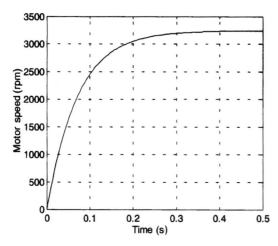

Figure 8.25 Motor Response to Step Voltage Input

$$\frac{d\Omega}{dt} = -a\Omega + bv \tag{8.26}$$

where

$$a = \frac{JR}{K_\tau^2} \quad b = \frac{K_\tau}{JR} \tag{8.27}$$

A common way of identifying these parameters is by exciting a motor that is initially stopped with a step change in voltage. Figure 8.25 shows a simulation of such a step response for a 24-V input. The motor parameters are typical of those of a small dc brush-type servomotor driving a purely inertial load.

   (a) Determine a (and thus τ, which is its reciprocal) and b from the graph. Do this by referring to the analytical solution given in Problem 2. Make sure to specify the units that are being used.
   (b) Determine a reasonable estimate of the uncertainty in this parameter determination by having several people perform the identification independently. In this case, since the data were generated from a simulation rather than from a real motor, all of the uncertainty comes from the process of processing the data rather than from the original experiment.
   (c) Set up a real motor with a substantial inertial load and do the experiment described above. Compare the results to the manufacturer's specifications for the motor.
   (d) Repeat part (c) several times to find the range of results that can be obtained and the repeatability of the experiment.

4. The model used to estimate motor parameters in Problem 3 ignored friction. The model can be enhanced to include simple Coulomb (dry) friction,

$$\frac{d\Omega}{dt} = -a\Omega + bv - f_F\frac{\Omega}{|\Omega|} \tag{8.28}$$

The ratio of motor speed divided by its absolute value always has a magnitude of 1 but changes sign as the speed changes sign [it may not be computationally determinant going through zero; most computer libraries include a signum function, sgn(·)]. Show that the parameters of this model can be determined from an experiment consisting of two step responses of different sizes.

   (a) Redo the experiments of Problem 3 with this model.
   (b) Experiment with different-size step pairs to see if the friction value obtained depends on the particular pairs chosen.

5. Use an analog tachometer and a fast data acquisition system or a digital oscilloscope to measure the velocity of a dc motor that has a constant voltage input (this is the same setup as that used for the stepping motor measurements). Observe the velocity variation over a range of input voltages. This should be done with a bare-shafted motor and with a motor having an inertial load.

6. Add a microcontroller having analog input and output channels or a laboratory computer with analog input and output to the setup of Problem 5 so that a velocity feedback control can be implemented.

   (a) Design and build the control software, then redo the tests of Problem 5 and see if there is any change in results.
   (b) Change to an eccentric load and compare open-loop (Problem 5) and closed-loop control of velocity.
   (c) Experiment with the sample interval. Plot performance (maximum or averaged error) versus sample interval.

7. Set up a motor with both position- and velocity-measuring devices and a microcontroller or computer that can read both and produce an analog output.

   (a) Write software to derive the velocity from the position measurement (be careful of scaling and units). Run the motor with constant input voltage and compare the estimate of velocity from position to the measured velocity. Change the input voltage to get a variety of speeds.
   (b) Compare feedback velocity control using measured velocity and estimated velocity.
   (c) Set up software for a cascade control to regulate position. Compare performance using the estimated velocity for the inner loop with that using the measured velocity.

9

Analog ⟷ Digital Conversion

Although digital devices are becoming dominant for signal processing and control, the physical world remains analog. All interfaces between computers and the physical world must thus contain some form of analog-to-digital conversion. In an increasing number of cases, the conversion is built into the device, actuation, or instrumentation. The stepping motor is a good example. Its command is digital, but its motion is analog. Instruments based on pulse generation, such as encoders, take analog motion as inputs and produce electrical outputs that are digital.

There remain, however, many actuators and instruments that are analog on both sides, taking a physical quantity as input and producing an analog voltage (or current) as output, or vice versa. The dc motor and dc tachometer are good examples of devices in this class. In these cases the necessary analog-to-digital conversions are done by devices that become part of the computer system. Because the conversion involves electrical media on both sides, that is, analog electrical on one side and digital electrical on the other, regardless of the physical media originally involved, a large number of purely electrical devices for conversion from analog to digital and digital to analog are available. An analog-to-digital converter (A/D or ADC) takes analog information and turns it into a digital word that can be accessed by computer programs (or logic circuits). A digital-to-analog converter (D/A or DAC) takes a digital word and converts it to an analog signal.

9.1 CODING

The major distinction between the analog and digital domains is in the way that precision is defined. Analog quantities can take on any values inside their range of validity. The precision with which the value can be read depends only on the instrument doing the reading. There is no inherent limit in the precision. What limits the actual precision, though, is the amount of noise present in the signal. Regardless of the precision of the instrument being used, the *measurement precision* is no better than the signal/noise ratio. Noise is defined as the component of a signal that is independent of the information being transmitted by the signal. If the source of the noise and its characteristics are not known, it is impossible to separate the signal from the noise, so any measurement made can be known only to a precision determined by the noise magnitude.

If something is known about the noise, some separation can be attempted. The most common separation is on the basis of spectral content. It is often known, for example, that most of the spectral energy in the noise is above some frequency, whereas the signal of interest lies below that frequency. In that case, filtering can be done to separate signal from noise and thus increase the precision in the resulting signal. In this specific example, a simple low-pass filter could be used to attenuate noise while passing the signals of interest without attenuation.

The signal/noise ratio for a given system is usually quoted in terms of the full-scale value of the signal. As long as the signal is near the limits of its range, the precision is well represented by the signal/noise ratio. When the signal gets small, however, the noise remains the same, so the signal/noise value deteriorates. The actual precision in a measured quantity is thus usually dependent on its magnitude. Digital signals, on the other hand, are *quantized*. That is, they can take on only a finite number of values. Binary (base 2) numbers are used in most cases, because digital computers are based on binary (Boolean) components. In some cases, though, decimal (base 10) numbers are used, particularly in devices that also contain readouts for operator use. Binary digital values will be assumed in this discussion.

The inherent precision limit in a digital signal is determined by its quantization. An eight-bit digital number, for example, can have 256 unique values. Noise is also a precision limit but only if the noise level is higher than the quantization level. The coding scheme used to represent the relationship between the digital and analog quantities is a design decision made by the manufacturer of the analog-to-digital device. There are several common codes, and it is not unusual to have to deal with more than one code in the same system. The simplest code is the representation of a unipolar analog signal with an unsigned binary integer. In the eight-bit example above, the digital numbers would be the digits 0 to 255 (00000000 to 11111111). The equivalences are easier to see using fewer bits, so in Table 9.1, a 0- to 5- V signal is shown represented by a three-bit digital signal, with 0 V corresponding to 000 and 5 V corresponding to 111. For this three-bit, 5-V converter, the quantization step is 5/7 V. In general, for an n-bit converter the quantization step is

$$Q_{\text{step}} = \frac{V_{\text{fullscale}}}{2^n - 1} \tag{9.1}$$

TABLE 9.1   UNIPOLAR CODING

Voltage	Digital value	Decimal equivalent
0	000	0
0.71	001	1
1.43	010	2
2.14	011	3
2.86	100	4
3.57	101	5
4.29	110	6
5	111	7

As seen from Table 9.1, the quantization error can be up to the size of a full step (usually referred to as the *least significant bit*, LSB). This can be reduced to half of the least significant bit by shifting the values to a centered position. This would give Table 9.2. When the voltage is bipolar, taking on negative as well as positive values, there are two codings that are used most often, 2's complement and offset binary.

TABLE 9.2   UNIPOLAR CODING WITH CENTERING

Voltage	Digital value	Decimal equivalent
0	000	0
0.36	001	1
1.07	010	2
1.79	011	3
2.50	100	4
3.21	101	5
3.93	110	6
4.64	111	7

Two's complement is the coding that is generally used for representing signed integers in computers. To review the counting material from the Boolean logic section, the positive integers count up from zero, as normal. The negative integers count down from zero, with the "borrow" discarded, so, in a three-bit system, -1 is 000 - 001 = 111. There are only eight unique values, including zero, so there is room for four numbers of one sign and three of the other (or four if zero is considered positive). The convention in 2's complement is to assign all of those combinations with the highest bit 1 to the negatives. Counting up from zero, the signed integers are 000, 001, 010, and 011 for 0, 1, 2, and 3. Counting down, the negative integers are: 111, 110, 101, and 100 for -1, -2, -3, and -4. The logical coding for analog values is to put analog zero at 000. With the asymmetric structure, the highest positive and negative values can either be made symmetric, thereby wasting one code, or made asymmetric, giving a different range for negative voltages than for positive voltages.

In offset binary coding, the bipolar voltage is scaled onto a set of unsigned integers. If the largest negative voltage is placed at 0 and the largest positive voltage at the highest possible number (all 1's), then because there are an even number of digital values, zero voltage will fall between the two middle digital values. Again, the same assignment choices made in the 2's-complement case have to be made here.

Offset binary has two advantages, which make it popular. The first is that the circuitry is usually a little simpler. The second is that the issue of *sign extension* does not have to be faced. Sign extension must take place whenever a 2's-complement number is converted from one precision environment to another. For example, if a 2's-complement number is represented in an eight-bit environment, -1 is 11111111. If that number is converted to a 16-bit environment, -1 is 1111111111111111. Note that the upper eight bits are all 1's. If a positive number is converted, +1, for example, 00000001 becomes 0000000000000001. The upper eight bits are filled in with 0's. All negative numbers have their highest bit set to 1 and must be filled with 1's when going to a wider environment.

Because the precision used by analog-to-digital converters is independent of the number precision being used by the computer, sign extension can be inconvenient for a converter and can require use of extra bits in the interface for the sign extension. If that is not done, sign extension becomes a software task. Two's complement's main advantage is that it matches the format used for negative numbers in most computers. Tables 9.3 and 9.4 give possible codings for a – 5- to +5-V signal using a three-bit digital value. The symmetric offset binary coding is used, so there is no "true zero" value. This seems extreme with three-bit conversion but is much less of a problem when higher-precision digital values are used. Zero is placed at 000 in the 2's-complement coding, giving one unused code.

TABLE 9.3   OFFSET BINARY CODING

Voltage	Digital value	Decimal equivalent
–5	000	0
–3.57	001	1
–2.14	010	2
–0.71	011	3
+0.71	100	4
+2.14	101	5
+3.57	110	6
+5	111	7

TABLE 9.4   TWO'S-COMPLEMENT CODING

Voltage	Digital value	Decimal equivalent
—	100	–4
–5	101	–3
–3.33	110	–2
–1.67	111	–1
0	000	0
+1.67	001	1
+3.33	010	2
+5	011	3

9.2 DIGITAL-TO-ANALOG CONVERSION

A digital-to-analog converter (also called a D/A or a DAC) allows a digital value stored in a computer to be output as an analog voltage or current. Any of the codings described above can be used for the conversion. Although the output voltage will, ideally, only take on the discrete values associated with the digital inputs, it is still an analog quantity. Unlike a digital quantity, small changes in its value due to deviations from ideal behavior (e.g., circuit loading, misadjustment, etc.) will cause changes in the system behavior.

A D/A converter is shown in block form in Figure 9.1. The digital value is transmitted from the computer to a latch (memory register). The latch will keep its last value until a new value is sent to it. The latch is connected to the actual converter, which outputs the voltage equivalent to the digital number stored in the latch. (The output could also be in the form of an analog current, although voltage will be referred to here.) Because of the action of the latch, the analog voltage will be "held" at a constant value until the next time the computer changes its value. This generates the familiar staircase pattern observed at the output of the D/A converter. A device with this characteristic is also known as a *zero-order hold*.

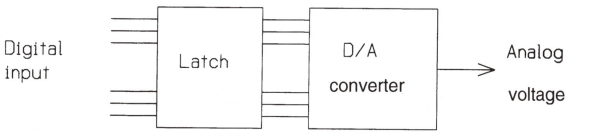

Figure 9.1 D/A Converter and Latch

Figure 9.2 shows a generalization of a three-bit D/A converter circuit. It makes use of analog switches, constant multipliers, and a summer. The analog switch has a digital input that controls whether or not the analog input will be transmitted to the output. If the switch is off (0), nothing is transmitted; if it is on (1), the full value is transmitted. The constant multipliers are set up to match the binary place values of the individual digital signals, and the final result is summed to get the analog output. This block diagram is a direct translation of the definition of a base 2 binary number as realized by a set of Boolean signals. It can be implemented in this form using an operational amplifier circuit (see Chapter 11), but for more than three or four bits the resistor values needed cover too wide a range. A more useful form of the circuit, which only requires resistors of two values regardless of the number of bits, is the ladder circuit, also shown in Chapter 11.

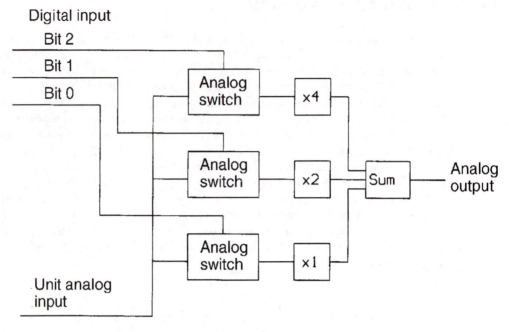

Figure 9.2 Power-of-2 D/A Converter

Inaccuracies in D/A converters come primarily from variations in the reference volt-age (marked in the figure as the "unit analog input") and changes or inaccuracies in the resistors. In addition to manufacturing variances, both of these quantities are subject to drifting because of temperature change. The precision of the converter is determined by the number of digital bits in its input, but its overall accuracy is determined by these circuit factors. The only speed limitation in the operation of these types of D/A converters is the operating speed of the circuit components. It is normal to expect that the output voltage will settle to its new value within a few microseconds in most standard converters.

9.3 FLASH ANALOG-TO-DIGITAL CONVERSION

Analog-to-digital conversion is considerably more complicated than digital-to-analog conversion. Four different technologies for A/D conversion are examined here. The first, flash conversion, is the fastest, but, usually, has the least precision. Flash conversion is carried out by a set of simultaneous voltage comparisons. This can be illustrated with the example of a two-bit converter for a voltage in the range 0 to 5 V. The goal of the converter is to produce a digital ouput that is representative of the analog voltage presented to it. The digital output is coded as unsigned integers, corresponding to voltage ranges of the input according to the code of Table 9.5.

TABLE 9.5 CODE FOR 2-BIT A/D CONVERTER

Voltage range	Digital value
0–1.25	00
1.25–2.5	01
2.5–3.75	10
3.75–5	11

Three comparisons can determine which of the zones the voltage falls in. The three comparators, C1, C2, and C3, are defined as follows:

C1 $V > 1.25$
C2 $V > 2.5$
C3 $V > 3.75$

Each of the variables C1, C2, and C3 is a Boolean variable. The converter output can be represented by a truth table, with C1, C2, and C3 and inputs, and the high (H) and low (L) bits of the output as output (Table 9.6). The don't-care outputs are for the combinations that "should not" occur. If the system is operating properly, it is impossible for C3 to be 1 when either C1 or C2 is 0, for example.

TABLE 9.6 A/D CONVERTER TRUTH TABLE

C3 C2 C1	H L
000	00
001	01
010	—
011	10
100	—
101	—
110	—
111	11

With this information a hybrid circuit, part digital, part analog, can be built to make the converter. The analog part consists of the set of comparators, which must have digital outputs compatible with whatever logic family is being used. The digital part implements standard logic. This type of converter can be very fast, as fast as the circuit elements can operate, which is why they are called *flash converters*. Commercially available units can have conversion times of 50 ns or less.

They do have a serious problem, however. Since comparison for all possible outputs must be done at the same time, the number of comparators will go up with the power of the precision. A two-bit converter, as shown above, is simple to implement, but going to three-bit precision will require seven comparators instead of three, and an eight-bit version needs 255, plus enough logic circuitry to take 255 inputs and produce eight outputs. The comparators must, at a minimum, be consistent enough so that there is no range "crossover," which would give a nonmonotonic output. They are thus most useful for applications that require high speed but relatively low precision.

9.4 SUCCESSIVE-APPROXIMATION A/D

The workhorse of A/D converters is based on successively checking voltage levels of a partially converted result to the test voltage. An iterative procedure is used, starting with the highest-order bit. Successive-approximation A/D converters necessarily contain D/A converters within them. Unlike flash converters, conversions of higher precision require more time. The increase in components, however, is relatively modest when compared to flash converters. To achieve the higher precision, though, the components that are used must be of higher quality.

The D/A converter should be set so that it produces the boundary voltages of the ranges defined above. For this case, those voltages are as shown in Table 9.7. The first step, then, is to set the high bit to 1, giving a value of 10 for the first trial. When that is sent through the D/A converter it produces an output of 2.5 V. That is tested against the unknown voltage. If the unknown voltage is less than the D/A output, the conclusion is that the high bit should be a 0; otherwise, 1 is correct.

TABLE 9.7 D/A CONVERTER VALUES FOR TWO-BIT SUCCESSIVE APPROXIMATION A/D

Digital value	Voltage
00	0
01	1.25
10	2.5
11	3.75

Then the next bit is checked, the only other bit in this case. With the high bit set at its correct value, it is set to 1 and the same test is repeated. In this case, if the high bit were 1, the digital trial output would be 11, which would test the unknown voltage against 3.75 V. This would serve to determine if the correct range is 2.5 to 3.75 V or 3.75 to 5 V, since it is already known that the unknown voltage exceeds 2.5 V. On the other hand, if the high bit were 0, the digital trial value would be 01, so the test would be against 1.25V.

This entire procedure can be summarized with the iterative logic shown below. This could be implemented in software very easily, but the conversion rate would be very slow. It is normally implemented as a logic circuit, with conversion times from 1 to 100 microseconds for conversion precisions from 8 to 16 bits. A very popular precision for successive-approximation converters is 12 bits, which gives a least count of 1 part in 4096, or about 0.025% of full scale, which is better than most analog circuitry. Conversion precisions of 16 bits (and sometimes even more for particularly demanding applications) are available if needed.

Successive-approximation logic

```
Start Conversion
Result = 0
For i = n–1 to 0
    Set i–th bit of Result to 1
    If unknown voltage < DtoA (Result)
        Set i–th bit of Result to 0
End of loop
Output Result
```

The general form of an A/D converter is shown in Figure 9.3. Because the successive approximation-converter is relatively slow, the voltage could change during the time the converter was working. To avoid this, a *sample-and-hold* unit is often placed before the converter. This is an analog device, which can capture an analog value and store it. It has a very short window time compared to the conversion time, so establishes unambiguously the moment at which the converted voltage is sampled. The storage in a sample-and-hold device is usually capacitive, so if there are frequencies even higher than it can capture, the signal will be averaged over the window time.

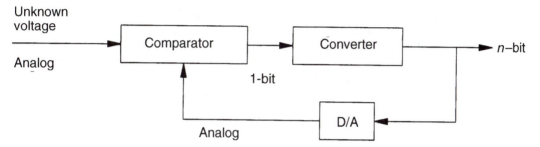

Figure 9.3 General Configuration for Successive-Approximation A/D

A/D converters are relatively expensive and for many systems can convert at a faster rate than is required for an individual channel. To allow conversion on several channels to share the same converter, an analog *multiplexer* is often used. This acts as the equivalent of a rotary switch, connecting the desired channel to the converter on command. Electronic multiplexers can switch channels in a few microseconds, so do not add appreciably to the operating time of a converter. If very low level signals are being switched, such as those from thermocouples, the impedance of electronic multiplexers is too high, so multiplexers based on relays must be used. These are much slower, with switching times in milliseconds. The primary error source in successive-approximation converters is the D/A converter. Any drift or change in its characteristics will cause errors in the output.

The computer interface to a succesive-approximation converter normally utilizes a *start* signal and some form of flag to indicate when conversion is complete. In systems that are operating with very tight timing constraints, it is possible to connect that flag to the interrupt system. The converters are fast enough, however, that the interrupt latency is a substantial fraction of the convert time, so using interrupts between the start of conversion and the end of conversion is not very efficient. If the conversion is based on time, it is often possible to set the converter up so that the conversion is started as a clock runs out, and an interrupt is generated only when the conversion is finished. This can work very efficiently. When very fast conversion rates are needed, at or near the maximum conversion rate, a direct memory access (DMA) control can be used. When a conversion is ready, the converter will take control of the memory bus and transfer the result directly to memory. When an entire block has been transferred, an interrupt is generated.

9.5 INTEGRATING CONVERTERS

Integrating A/D converters are very slow (as much as several hundred milliseconds per conversion), but are the most cost-effective way to produce very precise and very accurate results. In simplest form, their accuracy characteristics are the same as those of successive-approximation converters. They operate by generating a signal that starts at the low end of the voltage range and ramps (changes at a constant rate) up at a known rate. That signal is fed to a comparator that has the unknown voltage as its other input, (Figure 9.4.) The comparator switches when the ramping voltage crosses the unknown voltage, stopping the clock, which had been started at the same time as the ramp. The clock now contains a number proportional to the unknown voltage. The time period measured is the time it took the ramp to go from zero (or the bottom of the range) to the unknown voltage. The precision of an integrating converter depends on the amount of time allowed for the conversion, and the precision of the components used, particularly the ramp generator. Any desired precision can be achieved if enough time is available.

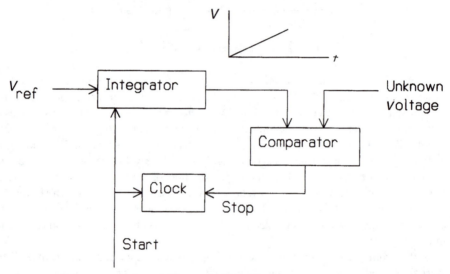

Figure 9.4 Integrating A/D Converter

The accuracy of such a converter depends on the same kinds of factors as the successive approximation and flash converters. However, much more accurate converters can be made using integration techniques if more time is allowed for the conversion by using multiple-slope conversions. In a *dual-slope converter*, for example, the unknown voltage is integrated up for a fixed period, then the time necessary to integrate back to zero is measured. This form of conversion cancels the effects of several of the circuit parameters, giving a more accurate conversion. Higher-order conversions, to make the converter even more independent of parameter values, temperature, and so on, are also possible.

9.6 SIGMA-DELTA ONE-BIT CONVERTERS

Sigma–delta converters are difficult to categorize. For starters, they are symmetric in the sense that the same structure can be used to create either an A/D or a D/A converter. In the speed range, they usually overlap with successive-approximation converters. In operating principle, however, they look most like pulse-width modulation. Developments in sigma–delta converters have been driven by the need for low-cost, high-precision converters. The lowest cost can be achieved in high-production devices by incorporating the converter onto the same integrated circuit as the other functions needed for system operation. Since converters by nature have both analog and digital information, the IC chips that include them must be *hybrid* designs (i.e., have a mixture of analog and digital circuitry on the same chip). There is a strong premium in hybrid chips to minimize the amount of analog circuitry since it is more difficult to manufacture reliably and also occupies more space than digital circuitry. The sigma–delta converter fills this need by increasing the reliance on digital circuits and minimizing both the amount and the sensitivity of analog circuitry needed for a given conversion performance.

The best known example of sigma–delta converters is in compact-disk players that are advertised as having "one-bit" converters. These are sigma–delta D/A converters. In broad terms, they have no advantage over other types of D/A converters, except for lower cost, but the term has been seized on by the advertising industry as a distinguishing characteristic.

9.6.1 Delta Modulation

The delta modulator serves as the basis for sigma–delta converters. The key idea behind this modulation scheme is to "track" the incoming analog signal in an average sense and produce a digital waveform that can be used to represent the analog signal. The demodulation process recovers an analog signal from the digital waveform by filtering. As with any modulated signal, the modulation (or carrier) frequency must be much higher than the highest frequencies of interest in the original signal in order for the representation to have a high degree of fidelity.

Figure 9.5 on the next page shows a delta modulator and a delta demodulator. As shown, the output of the modulator, $d(t)$, is fed directly into the demodulator. This is done to show how the modulated signal is created and interpreted; in practical applications, some form of digital processing would normally take place on $d(t)$ before putting it through the demodulator. The output of the modulator, $d(t)$, is a digital signal, since it is the output of a comparator. The input signal, V_{in}, and the internal (estimated) signal, V_{est}, are analog signals. By summing the modulated signal through the integrator an estimate of the input is produced. The feedback comparison then produces an error signal that is then used to create $d(t)$. As the sign of the error changes, $d(t)$ switches from one digital value to the other. The carrier frequency of this process is determined by the clock frequency used for the clocked comparator.

Delta modulator

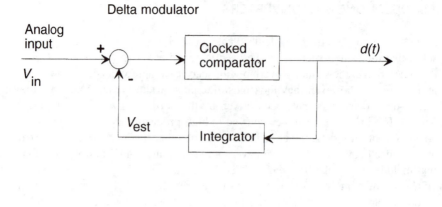

Delta demodulator

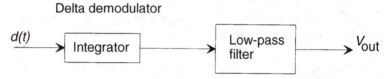

Figure 9.5 Delta Modulator/Demodulator

The top graph of Figure 9.6 shows the input (solid) and estimated (dashed) signals; the middle graph shows the modulated signal, $d(t)$. The comparator used outputs of -1 and +1. A lower-than-normal clock frequency is used so that the nature of the signals can be seen clearly on the graph. In practice, the clock frequency must be high enough so that the error in V_{est} is within the project specifications. The demodulation process first passes the modulated signal through an integrater, then filters out the high-frequency (carrier) part of the signal in a low-pass filter. The input (solid) and the demodulated signal (dashed), V_{out}, is shown in the lower graph of Figure 9.6. The phase and amplitude differences are due to the action of the low-pass filter, which was only first order for this simulation.

9.6.2 Sigma–Delta Modulation

Sigma–delta converters use a feedback structure similar to that of the delta modulator to produce a modulated one-bit digital signal at the output of the comparator (Figure 9.7). This same basic structure applies to both analog-to-digital and digital-to-analog convert-ers. The modulation process, which is a generalization of the delta modulator shown in Figure 9.6 is the heart of the converter and the reason why so little analog circuitry is needed. The output of the converter is at the output of the filter. This block diagram shows generalized signal domains, so it can represent either an A/D or a D/A converter.

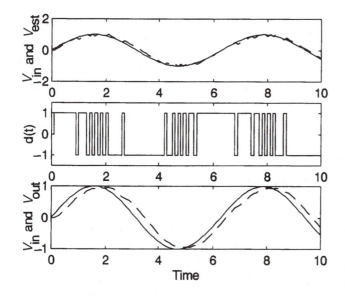

Figure 9.6 Simulation of Delta Modulation/Demodulation

Three scale domains are shown in the figure. The V-scale domain indicates signals in the same units as the input signal, V_{in}; the one-bit domain consists of signals that are one-bit digital values; and the F-scale domain is scaled to the output units. Note that, as pictured, this is a generic sigma–delta converter — it can be implemented as either a D/A or an A/D converter. The feedback loop is constructed in such a way that *on average* the output of the scaler will be equal to the input value, V_{in}. The input to the scaler is a one-bit signal, so its output will be either the minimum or maximum of the V-scale range. If it is the high limit value, it will generally be larger than V_{in}, so the output of the subtraction will be negative. This causes the integrator to start integrating down; when its output crosses zero, the comparator switches, the scaler's output switches to the low V-scale limit, and the integrator starts integrating up. Since the *scaler* can only produce extreme-valued outputs, there will be constant switching going on in the comparator. The result is that the output of the comparator is a modulated one-bit signal such that its average value corresponds to the converter's input value. The filter does the averaging necessary to turn the modulated signal into a scaled representation of the input. This is illustrated below with an example, but first this generic converter will be turned into an A/D, then a D/A converter.

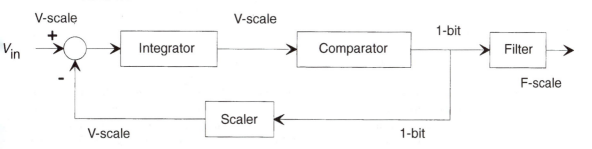

Figure 9.7 Sigma–Delta Converter Structure

This sigma–delta structure is specifically designed for use as a converter, as compared to the more general modulation structure of the delta modulator. As such, it has some advantages over the delta modulator shown in Section 9.6.1. First, it requires only one integrator. Second, the designer of the converter has more control of how quantization noise will influence the device and therefore, can tailor it to the intended application. Third, the sigma–delta structure is better at dealing with rapidly changing input signals.

9.6.3 Sigma–Delta Analog-to-Digital Converter

In an A/D converter the V-scale domain is analog. The input signal is the unknown analog voltage; the subtractor and integrator are analog components; the comparator is, in effect, a one-bit A/D converter; the scaler is a one-bit D/A converter; and the filter is a digital filter. Although the D/A converter is only one bit, it still plays an important role. Whereas its input is a digital signal whose actual voltage value can fall anywhere in the acceptable range for the logic family involved, its output is analog and must conform to the proper voltage within the noise tolerances on the analog side.

The F-scale domain is an n-bit digital domain, where the n-bit value is the output of the analog-to-digital conversion. The digital filter is often implemented with a digital signal processor (DSP). This structure is shown in a relabeled version of the converter structure (Figure 9.8). The one-bit domain is assumed to be synchronous, so that the operation of the one-bit ADC (comparator) and the digital filter are synchronized by a common clock.

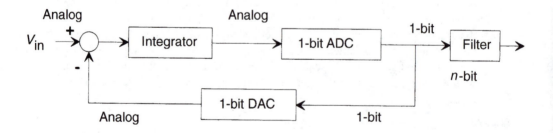

Figure 9.8 Sigma–Delta A/D Converter

An example of the output from a simple operating A/D converter is shown in Figure 9.9, which shows the output of the integrator and the comparator, and Figure 9.10, which shows the output of the filter. The input domain is analog, with a range of -1.0 to +1.0. The input signal is a constant voltage of 0.3. The output range is 0.0 to 1.0. The simple design, in both the converter and the digital filter, is used so that the operating principles can be demonstrated in a short simulated record.

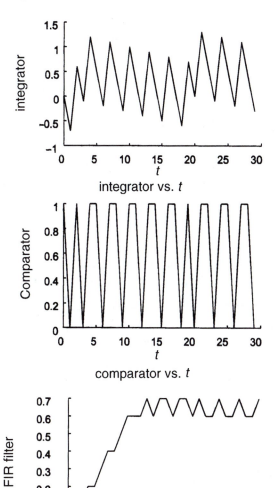

Figure 9.9 Sigma–Delta Converter
Operation: Integrator and Comparator

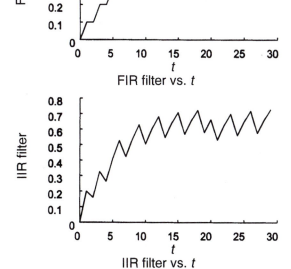

Figure 9.10 Sigma–Delta Operation:
Filter Output

Looking first at the comparator (one-bit ADC) output, it takes on only the digital values of 0 or 1 (the time scale is in number of time steps; sloped lines are interpolated from point to point by the drawing program). Consistent with the fact that the input value is in the top half of the input domain's range (1.3 out of 2.0), the comparator has a value of 1 for more of the time than the value is 0. The integrator, on the other hand, crosses back and forth across zero, causing the comparator to switch each time it crosses zero. Again, consistent with the input value, the integrator is above zero for more time than it is below. The integrator is in the V-scale domain and takes on arbitrary values between -1.0 and 1.0.

Output signals are shown for two types of filters, an FIR (finite impulse response) filter and an IIR (infinite impulse response) filter. The IIR filter is a digital version of a standard analog filter. In this case, the first-order filter implemented is described by the input–output relation,

$$y_k = (1 - C)y_{k-1} + Cu_k \tag{9.2}$$

where u is the input to the filter and y is its output. This type of filter has a theoretically infinite tail. That is, if u is a constant, y will approach that constant value exponentially. The resulting signal takes about 15 time steps to reach its final value, then varies around its final value with an amplitude of about 0.1. This variation is what determines the precision of a sigma-delta converter. In this case the variation is about 1 part in 20, giving a converter with about four-bit precision.

An FIR filter differs from an IIR filter in that the right-hand side of the filter equation has values of the input, u, and its past values, but the previous values of the output, y, do not appear,

$$y_k = C_o u_k + C_1 u_{k-1} + C_2 u_{k-2} + \dots \tag{9.3}$$

The simplest form of this filter is when all the coefficients have the same value, $1/n$, where n is the number of terms. This is also called a *boxcar filter*; it is a moving average of the most recent n points. The IIR filter shown in the figure is of the boxcar type, with $n = 10$. It has roughly the same rise time as the FIR filter and about the same amount of variation — but many more terms. There is a major advantage of using an FIR filter with a sigma–delta converter, however. Because the inputs are all 0's and 1's, no multiplications are necessary; if an input is 1, the corresponding coefficient is added into the sum. Otherwise, the term is skipped.

The FIR filter is also more sensitive to patterns in the input. In Figure 9.11, for example, all parameters are the same as in the previous case, except that the input, V_{in}, is zero. This is exactly at the midpoint of the V-scale range, so the modulation pattern is completed within the 10 samples of the filter, so no output variation appears after the filter has settled. In this case, the IIR filter shows its characteristic variation, but the FIR filter does not. Another advantage of the FIR-based sigma–delta converter is that it handles oversampling effectively (another advertising buzzword for CD players). This is described in more detail in Section 9.7.

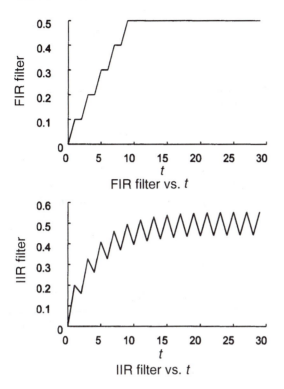

FIR filter vs. t

IIR filter vs. t

Figure 9.11 Filter Output for $V_{in} = 0$

9.6.4 Sigma–Delta Digital-to-Analog Converters

The D/A sigma–delta structure is shown in Figure 9.12. The input value is the digital number to be converted, the accumulator is a digital summation element, and the comparator and scaler are also fully digital components. The filter is an analog filter. Because it is analog, the filter will of necessity be of the IIR type. The performance curves given above can just as easily be interpreted as D/A performance as they were for A/D, except that the FIR filter outputs should be ignored since they cannot be realized in analog form. This system is a version of the one-bit DAC of CD fame.

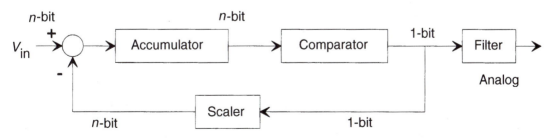

Figure 9.12   Sigma–Delta Digital-to-Analog Converter Structure

9.6.5 Sigma–Delta Design Considerations

Design of practical sigma–delta converters focuses on three aspects of the design: the modulator, the filter, and cost. The structures shown previously are first-order modulators. Higher-order modulators can be used to improve noise rejection and to increase the dynamic range of a converter. Multiorder feedback or feedforward structures can be used. The feedback structures are simpler but harder to design. Because of the feedback structure used for all sigma–delta converters, there are possibilities of instabilities caused from loop closure. It is more difficult to assure stability for higher-order feedback structures because the comparator and scaler are nonlinear elements and make the stability analysis difficult or impossible. Stability cannot be guaranteed for higher-than-second-order sigma–delta structures. Higher-order feedforward structures, on the other hand, have guaranteed stability since they use cascaded first-order elements that are guaranteed to be stable. However, they use more components for the same performance.

Filter design is a major course of study in itself. For sigma–delta DACs, the output filter is analog, so the design decisions involve standard filter design parameters. For the ADC, however, it is possible to use either FIR or IIR filter structures. IIR filters achieve similar performance with fewer terms (as in the previous example) but are more difficult to design. FIR filters also have linear phase response (which is of particular importance in audio applications) and can be used directly in decimation configurations (see Section 9.7.2).

Cost is often more of a consideration in sigma–delta converters than in other types because they are peculiarly applicable to incorporation in high-volume integrated circuits. Extremely careful specification of input and noise signal characteristics and realistic specification of performance constraints are critical to arriving at the most cost-effective design.

9.7 SAMPLING

Sampling is a fact of life when computers are used. With A/D converters, the data are taken only when the converter is activated. This inevitably involves a loss of information during the time when the signal is not being examined. The major design issue that must be addressed is the selection of circuit components and sampling frequencies in a way that will minimize the information loss.

9.7.1 Aliasing

Aliasing is the primary source of signal contamination in sampled systems and is uniquely a phenomenon of sampled systems. It occurs whenever there are elements in the signal at frequencies that are higher than half the sampling frequency. When the signal is sampled, these elements of the signal appear in the sampled data at full amplitude, but with their frequency transformed into the lower-frequency range of the sampled signal.

This process is easier to imagine with an extreme example. Assume a signal that is a pure sinusoid at a frequency of one cycle per unit time. If the signal is sampled relatively rapidly, 20 samples per unit time, for example, the sampled data signal will very closely resemble the original signal. In Figure 9.13 the top curve is the original curve and the second curve from the top is sampled at 20 samples per unit of time. As the sample rate is reduced, the resemblance will get cruder and cruder, but the basic characteristics will remain. Even at low sampling rates such as 5 samples per unit of time, the fact that the signal is periodic, its frequency, and its amplitude will all be readily discernible (see the third curve). If the sampling frequency drops still further, however, then even the periodic nature of the signal will no longer be obvious. At a sampling rate of, for example, 0.7 sample per unit of time, it is not even possible to represent a periodicity of 1 cycle per unit of time (see the bottom curve). The information content of the signal sampled is too low. The property of particular importance in the bottom curve is that while all information about the shape of the original curve has been lost, the full amplitude has been preserved. This is the source of the considerable mischief for which aliasing can be responsible.

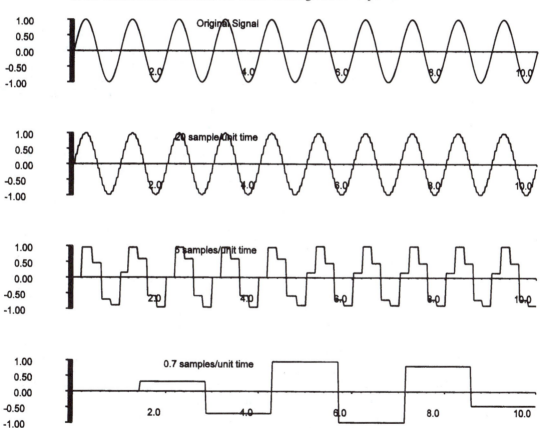

Figure 9.13 Sampling of a Sinusoidal SIgnal

In signals that have low-frequency information and high-frequency noise, the sampling rate is normally set to capture the information. As shown above, though, the noise will come along also, and at full amplitude. Its frequency will not be preserved, so the noise component will now have its frequency content in the same range as the information and will no longer be filterable. A signal with separable noise and information has thus been transformed by aliasing into a signal with just as much noise but no separability. The only solution to a serious aliasing problem is to place a low-pass analog filter upstream from the converter. This filter, an antialiasing filter, should be set so that it removes all frequency content above one-half the sampling frequency (the one-half limitation comes from the Nyquist and Shannon sampling theorems, which establish the theoretically lowest sampling rate as twice the highest-frequency component of a signal).

It would be nice to build the antialiasing filter into general-purpose converters. Unfortunately, the sampling period is normally under software control, and can change over very wide ranges, so there is no way to generalize the filter requirements. It is thus left to the designer of each application to assure that there is adequate aliasing protection.

9.7.2 Oversampling

If fast enough analog-to-digital conversion is available, it is possible to sample much faster than needed by the application and then use digital filtering to remove unwanted spectral components. This allows the antialiasing function to be broken into two stages: an analog filter for the highest-frequency components and a digital filter for the lower-frequency components. This adds substantial flexibility to the converter, since it is much easier to change the tuning of a digital filter than it is for an analog filter. It can also be a cost-savings measure; in highly integrated applications, in particular, it can be less expensive to implement a digital filter than an analog filter. An analog filter is still normally required, but it operates at a much higher frequency than when the analog filter is responsible for the full anti–aliasing job, so uses much smaller and less expensive components.

Advertising copy to the contrary, oversampling converters are not inherently any better than conventional converters. They may, under some circumstances, be less expensive to build, or more flexible to use, but they do not have any better performance. The stream of data produced from oversampling is much faster than is needed by the downstream computation, so it *subsamples* from the stream by removing every N_{th} data point (where N is the oversampling ratio). This means that $(N - 1)/N$ of the data points produced by the converter/filter are not used for anything. FIR digital filters have a unique property that makes this process more efficient, however. Reviewing equations (9.2) and (9.3), the operational distinction between IIR and FIR filters is that the right-hand side of the IIR filter contains terms dependent on both inputs and outputs, whereas the FIR filter only contains terms dependent on the input. With an FIR filter, whenever an output is needed, the appropriate weighted sum of past inputs is computed. If no output is needed, the current value of the filter input must be stored, but no computation is required. Therefore, the actual filter computing load is only $1/N$ of the computation load needed for full-rate sampling (N is the oversampling ratio). This process, which combines computation and subsampling, is called *decimation*. Not doing $(N - 1)/N$ of the computations does not in any way diminish the performance of the FIR filter. IIR filters, on the other hand, need past values of both the inputs and the outputs, so all the outputs must be produced and stored even if they are not used by the downstream process. This property makes FIR filters very attractive wherever oversampling is used.

Oversampling is often used with sigma–delta A/D converters. To achieve high-enough frequencies for the modulated output (high enough so that filtering time constants are not too slow for the process needing the converted data) the base sampling rate is often very high. Sigma–delta converters need the digital filter at the output stage in any case, so using it in a decimating mode does not add any component complexity to the converter.

9.8 PROBLEMS AND DISCUSSION TOPICS

1. Construct a software successive–approximation analog-to-digital converter by using a computer with a digital-to-analog converter and an external comparator that produces a logic output that can be read into the computer's digital I/O port (a standard circuit).

    (a) What is the conversion time as a function of the precision of the conversion?

    (b) If a precision voltmeter is available, evaluate the quality of the conversion across the full range of input voltages.

    (c) Explore the behavior of the converter when the input voltage is changing during the conversion period (this simple converter will not have a sample-and-hold circuit). Is the result always within the range of change? Is it biased toward one end or the other of the range of change? Is the result biased toward values appearing early in the conversion period or those appearing later?

2. Design and build a three-bit flash converter and an interface with a signal to start the conversion and a means to read the result into a computer. Measure (with a fast sampling oscilloscope) the speed of conversion.

3. Modify the three-bit converter of Problem 2 so that the range of conversion is very narrow. Use this as an experimental test bed to see how difficult it is to maintain monotonicity of conversion with flash converters and how linear the conversion is.

4. Design and build a three-bit successive approximation analog-to-digital converter based on a synchronous control circuit. Use standard components for the comparator and D/A converter.

5. Design a software-based, integrating A/D converter (single slope). This will require the same equipment as the software-based successive approximation converter (Problem 1). Is there any advantage to the integrating configuration?

6. High-speed data acquisition requires transfer speeds from the converter into the conputer's memory that are much faster than can be obtained from software control of the converter. If a converter with DMA (direct memory access) is available, program it to acquire a burst of data from a high-speed process. This involves both setting up the DMA transfer and figuring out an appropriate trigger to start the burst.

7. While sigma–delta converters take a lot of circuitry to build, a reasonable approximation of a sigma–delta digital-to-analog converter can be built from a computer, an analog filter, and software. The accumulator, comparator, and scaler are all implemented with software. The comparator output is sent out through a digital output port and then through the analog filter. The speed with which the loop can be accomplished is equivalent to the clock rate of an actual sigma–delta converter. A data table or direct computation can be used to generate the function being converted. The standard A/D converter can be used to read the final result back into the computer for direct comparison to the value being converted, or a D/A converter can be used to send out the value for comparison on the analog side.

    (a) Verify that the system works by converting constant values and checking the result with an oscilloscope or voltmeter.

    (b) Use an oscilloscope to examine the output of the analog filter to see how much ripple is in the signal from the original modulated signal.

    (c) Adjust the cutoff frequency of the analog filter so that the ripple is reduced to some reasonable value (say, 2% of full scale). What cutoff frequencies must be used to get the ripple to one-half or one-fourth of its nominal value?

    (d) With the filter adjusted, examine the ripple as a function of the magnitude of the constant value being converted.

    (e) Put in a varying input signal, for example, a sine wave. Explore the range of frequencies that can get through the analog filter without significant phase or magnitude change.

10

Position and Velocity Measurement

It's hard to imagine a mechanical system that does not use a measurement of velocity and/or position somewhere in the system. Many instances are obvious and may involve the major functioning of a machine; other cases are less central but no less important, such as the positioning of valve stems in a process control system. Position control long precedes electrically based control, and a variety of mechanical, pneumatic, hydraulic, and other position and velocity instruments exist. Electricity has proven to be the most effective working medium thus far, however, so the instruments described here all produce electrical outputs.

As in most technological transitions, the first articulations of the transition are tentative and far removed from the actual point of measurement. Intermediate conversion steps are introduced to allow older technology to interface with the newer. For example, in systems that were pneumatically controlled, the original pneumatic position sensor might remain in place, with a pneumatic/electric converter inserted into the system. As the new technology becomes dominant, use of instruments that perform the conversion more directly improve the overall system performance and reliability while decreasing complexity and cost.

10.1 ANALOG/DIGITAL: PRECISION, RANGE, AND ACCURACY

The collection of instruments described here shows a similar transition within the electrical medium, from analog to digital coding and processing of signals. Initially, most of the conversion was done directly at the processor, using an analog-to-digital converter. Instruments such as encoders, however, are inherently digital and demonstrate their prime advantages only within a digital environment. The primary advantage of digital coding for instrumentation, particularly for position measurement, appears in *dynamic range*. As discussed in the chapters on digital logic, the design of a digital system should be such that a single bit of information can be delivered essentially noise free. Imagine a system, for example, that must position an object with a precision of 1 micron (1 μm) and must be able to move the object a total distance of 1 m. That requires a dynamic range of 1 part in a million, a standard that is very difficult in an analog positioning system.

In a digital system, the design for minimum count is decoupled from the design for maximum range. The minimum count problem is usually closely coupled to the physics of the problem. In position measurement, it involves the generation of some sort of signal that is coupled to the motion of the object in question at the desired precision level. To provide a digital signal, that coupling must provide a noisefree indication that, in the case under consideration, 1 micron of movement was made. Once that has been done, the maximum range is entirely up to the designer. The *word size* used in processing the signal governs the maximum count. Using digital hardware (logic circuits), the word size is fixed by the amount of hardware used. A system to accumulate counts from such an instrument can be made, for example, using 32-bit registers. This would achieve a maximum unidirectional count of about 4 billion (or $\pm$ 2 billion), far more than is needed for the problem mentioned above.

If the count is being maintained with software (i.e., using a computer with some form of interface to the device generating the count signals), the natural (or most efficient) word size is determined by the computer hardware. At the expense of computing time, though, any desired word size can be maintained. In hardware as well, any desired range can be achieved, at the cost in that case of additional circuit elements. Analog systems, on the other hand, usually couple precision (least count) with range. For any given technology used, there is an inherent signal/noise ratio, which normally refers to the full-scale signal. Thus, since the least count that can be measured is limited by the noise, the dynamic range is determined by choice of technology. A signal/noise ratio of more than 1 million:1 is most uncommon in analog systems (i.e., the reliable detection of a 10-μV change in a signal with a range of 0 to 10V).

The advantages of analog measurement are in its more efficient coding and its better match to the objects being measured, which are normally analog in nature. A single wire can carry an uncoded analog signal, with information content equivalent to 8 to 12 bits, depending on the noise level. At a single instant, a single wire can carry only one uncoded digital signal, one bit of information. Multiple wires can be used for the full signal, or serial transmission can be used in which a time window is used to transmit the full signal. The grounding and noise shielding, however, must be handled much more carefully in the analog system than in the equivalent digital signal, so the cost *per wire* of the analog transmission link is more expensive.

Digital processing is becoming dominant in mechanical systems. The first motivation for this is the ability to use computers for signal processing, which offer much more flexibility than their analog counterparts. The dynamic range arguments that relate to the measurement aspects of mechanical system problems are also a driving force. Processing speed has been the major impediment in this transition. Mechanical systems can require high processing bandwidths, which are achieved more easily with analog systems. Advances in digital electronics are narrowing that gap continuously.

10.1.1 Analog Velocity Measurement

A *tachometer* produces a signal that is related to velocity, usually rotational velocity. An analog tachometer produces a voltage proportional to rotational velocity. It uses exactly the same operating principles as a dc motor, brush or brushless, but runs as an electrically unloaded generator. The primary functional parameter is the back-emf coefficient, renamed the voltage constant, which relates the output voltage to the speed,

$$V_T = K_v \Omega \tag{10.1}$$

The other half of the motor relationship, torque to current, is irrelevant to a tachometer when used normally. On the electrical side, the output is usually fed into a device with a high input impedance, so the current drawn is negligible. On the mechanical side also, there is no significant torque load borne by the tachometer. Equation (10.1) is the first-order model for the operation of a tachometer.

10.1.2 Noise and Precision

Brush noise and switching transients represent the main noise sources in analog tachometers. Brushless tachometers do not have any brush noise, but they do have switching transients relating to the electronic commutation. Tachometers can be designed specifically to reduce the noise level and improve the linearity of the voltage–speed relationship. Because there is very little current flow through the commutator compared to the very large current flow the commutator must handle in a motor or generator application, its design can be based on noise reduction as the primary performance criterion. Ordinary small dc motors can be used as tachometers, but their performance is not as good as that of devices designed specifically for use as tachometers.

The low-velocity performance of a tachometer is limited by its precision specification, which is based on noise. At very low velocities, the signal gets to be smaller than the noise, so it is indistinguishable from noise. The actual resolvable velocity also depends on the characteristics of the noise and the ability to filter the incoming signal, reducing the noise level without affecting the velocity signal. The filtration is usually based on having a separation in the frequency characteristics of the signal and the noise, so is strongly problem dependent.

The low-velocity behavior is an important characteristic of many mechanical systems. There are two situations requiring good low-velocity resolution. The first is in determining whether or not a system is *stopped*. *Not moving* is a property that has a very strong intuitive definition, but because of noise and precision characteristics of velocity-measuring instruments, it is very difficult to relate the intuitive notion of stopped to a control definition. The most common solutions are to wait a specified period after the lowest measurable velocity has been attained, or to use a brake, which is applied when a suitably low (but measurable) velocity is reached.

The mechanical concept of stopping associated with a truly zero velocity is possible because of the characteristics of dry friction, which include a *sticking* effect that comes into play at very low velocity (motion is still possible, due to bending of the object; thus the definition of *stopped* as commonly used refers to relative motions). The name *stiction* is often applied to this phenomenon, standing for "stick-friction." In systems with significant stiction, the first method works because once the velocity hits the sticking level, the system will stop. Brakes depend on the same phenomenon and are used to increase the amount of dry friction.

The second situation requiring good low-velocity performance is in *scanning*. This operating mode uses constant velocity to perform an operation such as cutting, painting, welding, or inspecting. In many cases the velocity used for such an operation is much lower than the velocity needed to move the object from one working position to the next, the *slewing* velocity. In this case the dynamic range again becomes important, but for velocity rather than for position. A tachometer scaled for the slewing velocity might have too much noise for a successful scanning operation.

10.1.3 Range

The range limitation on the low end, as noted above, depends on noise. At the high end, the range is limited by the voltage capacity of the instruments receiving the tachometer's signal. Short of physical destruction, the tachometer has no practical voltage limit. It generates little or no current, so has no significant heating problem, which acts as a major limit for a motor or generator. Typical signal-level electronic components can handle up to about 10-V signals. The tachometer output should be limited if its physically possible voltage output is large enough to damage the electronics. Once the signal has saturated, the only information present is that the velocity is higher than the saturation value. In a scanning type of application this might be adequate if some other control mode is used to control slewing.

10.1.4 Velocity Control

The tachometer is the heart of the analog velocity control loop. This is the traditional loop that is used to provide stability to motor control systems, whether the ultimate control goal is velocity or position. It has traditionally been implemented using operational amplifier circuits, which gives a completely analog system. As digital control systems become more prevalent, variants on this theme are appearing. In the simplest variation, which puts the least demand on the digital portion of the control, the analog velocity loop remains in place and the digital section is used to implement the position control loop. After implementing the position loop with only a digital processor, the next step in some systems is to leave the tachometer in place but to use an analog-to-digital converter to read the tachometer output and a digital-to-analog converter to transmit the command signal to the amplifier.

The main purpose of the velocity control loop is to compensate for the dynamic effects of the motor itself and the coil inductance. A lead–lag element was used for this compensation in the example in the discussion of dc motor control. It is very common in velocity control loops, so is examined in more detail here. The basic description can be given in terms of a transfer function as

$$G(s) = \frac{T_1 s + 1}{T_2 s + 1} \tag{10.2}$$

To get a differential equation from this, the transfer function can be written in expanded form as

$$G(s) = \frac{T_1}{T_2} + \left(1 - \frac{T_1}{T_2}\right) \frac{1}{T_2 s + 1} \tag{10.3}$$

Introducing $\mathbf{x}$ as a variable, this can be converted directly to the differential equation,

$$\frac{dx}{dt} = \frac{1}{T_2}(u - x) \tag{10.4}$$

and the algebraic equation,

$$y = \frac{T_1}{T_2} u + \left(1 - \frac{T_1}{T_2}\right) x \tag{10.5}$$

In these last two equations, u is the input, y the output, and x an internal (state) variable.

This compensator is called lead–lag because it can take on either characteristic, depending on the values of its parameters. Figure 10.1 on the next page shows the behavior for two different parameter sets: the top curve corresponds to $T_1 = 1$ and $T_2 = 0.5$, while the bottom curve corresponds to $T_1 = 0.5$ and $T_2 = 1$. The input is a unit step (i.e., $u=1$ for all positive values of time). The top curve exhibits *lead* characteristics which are used for stabilization. The output, responds very quickly to the input, then dies away to a value equal to the input. This fast reaction is what caused the improved performance in the example in the dc motor example; the initial fast response tends to compensate for the sluggish behavior of the coil as the coil current comes up to its steady value in response to changes in voltage output from the power amplifier.

A simulation of a continuous-time (analog) velocity control is shown in the top graph of Figure 10.2. This uses a lead compensator and is similar to the example shown in Section 8.3.3 except that different coefficient values are used to account for saturation limits in the power amplifier (maximum voltage-handling capability) and the time scale is expanded. The need for fast sampling is related to stability of the (digital) control loop. To achieve the speed and quality of control desired, it is desirable for the controller to have high gains, that is, high values for the coefficients that relate the error (difference between desired and actual velocity) to the actuator command. As these gains are made higher and higher, however, the control performance will start to deteriorate, and finally, the loop will become unstable and oscillate uncontrollably. Slow sampling in the controller severely limits how high the gains can be before stability problems are encountered.

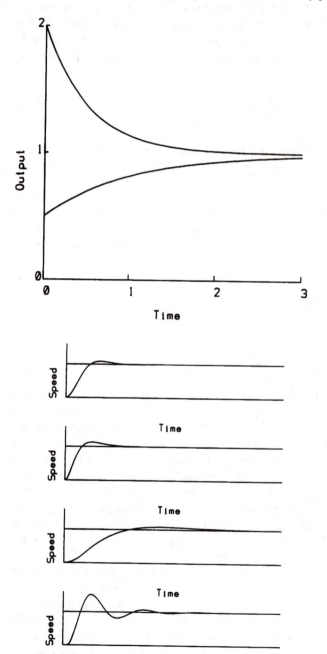

Figure 10.1   Lead–Lag Compensator Response

Figure 10.2 Analog and Digital Velocity Control

The lower three graphs in Figure 10.2 are all examples of digital velocity control. The second graph uses a relatively fast sample time, and the performance achieved is essentially the same as that achieved by the analog control. In implementing the simulation of this case, it was assumed that an instrument giving an instantaneous velocity measurement (such as a tachometer) was used and that the computer can calculate the controller output in a small fraction of the sampling interval (i.e., an "infinitely" fast computer). The third graph from the top uses a longer sampling interval but with the same instrument and computational speed assumptions. In this case, the controller gains must be reduced to avoid oscillatory or unstable behavior. The result is a slower response; it takes longer for the velocity to reach its new set point.

The bottom graph uses the same sampling time and controller gains as those of the second graph. However, instead of having an infinitely fast computer and an instantaneous measurement of velocity, it is assumed that the delays associated with computing and getting the velocity measurement amount to about the same time as the sample duration. Thus the new command signal (to the power amplifier) is delivered at the end of the sample interval, just before the start of the next sample instead of at the beginning of the sample interval as in the second and third graphs. The response in this case shows substantial overshoot and oscillation, even though the gains are exactly the same as for the second graph. The extra delay is a strong destabilizing influence.

10.2 PULSE MEASUREMENT OF VELOCITY

An early concept in digital measurement of velocity is the use of a device that is periodically excited to generate a series of pulses. If the excitation of the device comes from a moving (often rotating) object, the frequency of the pulse train is related to the speed of the device. This can be viewed as a direct digital measurement, because the signal produced is always in either an ON or an OFF state; no other information is inferred from the value of the signal. An alternative interpretation is that the signal thus produced is a modulated analog signal, with the modulation in the form of pulse frequency modulation (PFM). Either interpretation makes sense, and, to some extent, is resolved only when the signal and the downstream signal processing elements are considered together. Because of this ambiguity, however, pulse-generating devices represent a transitional stage between the analog and digital worlds.

10.2.1 Pulse Generation Devices

For devices that always move in one direction, or those for which the direction can be inferred from other information, a single pulse-generating channel can be used. The pulse can be created by a number of devices, including optical (reflective or beam-breaking), magnetic, pressure, reluctance, and mechanical switch. The shape of the pulse is not of critical importance. Sometimes a short pulse is created (relative to its off-time), and sometimes the device is arranged to produce square waves. The raw shape of the pulse itself will sometimes have a sharp rise and fall or may have a more rounded shape. The pulse should be reasonably clean of noise, however, to avoid false triggering.

10.2.2 Frequency-to-Voltage Converters

A frequency measurement will give a value proportional to the speed. This measurement can be made with a frequency-to-voltage converter. The output is an analog voltage that can be used with either an analog control system or, in conjunction with an analog-to-digital converter, a digital control system.

10.2.3 Logic Measurement of Frequency

A logic circuit to measure frequency needs a timer, a counter, and a few control circuits. The idea is to measure the number of pulses that occur within a specified period. The timer is used to control the time period, and the counter counts the pulses. A transition diagram for the overall control logic is shown in Figure 10.3. The output is held in a latch (register), which is updated at the end of each time period (during the read-counter state). The value latched is the number of pulses counted for the time period that was set. It is proportional to velocity if that time period never changes; otherwise, it must be divided by the time period to get the velocity.

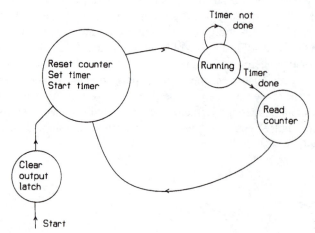

Figure 10.3 Velocity Measurement Logic

Two important safeguards could be added to reduce the possibilities for errors in this system. The first is to add a handshake with the "user" device to read the output latch. That latch is changed by reading a whole new value into it, which is a multibit change. Since all the bits cannot change at exactly the same time, if the latch is read while it is being changed, an erroneous result could be obtained. The handshake would assure that the latch is read only when it has valid data. The second change would cover the case when the counter overflows. A counter will normally roll over in those circumstances, producing a zero as the next count after a full count. If overflow should occur, the velocity read would be disastrously in error, indicating a very low velocity instead of a very high velocity. A second transition off the running state can check for overflow and either set an overflow flag or set the output to its highest possible value (Figure 10.4).

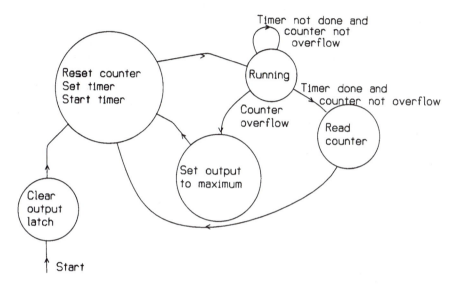

Figure 10.4 Velocity Measurement Logic with Overflow Check

10.2.4 Software Measurement of Frequency

These logic transition diagrams can also work for software measurement of frequency. If a processor is dedicated just to measuring velocity, the transition diagram can be the basis for a synchronous program. The timing would still normally be done with a hardware timer, since timing via program loops is unreliable and difficult to program. In more normal circumstances, the processor will have to be carrying on other control activities as well as measuring velocity. An asynchronous version of this system will require one or two interrupts, one for the timer and one for the pulse input, if no hardware counter is being used. If a hardware counter is used, the pulse input would be connected directly to the counter and would be read when the timer interrupt occurs.

10.2.5 Precision in Pulse Measurement of Frequency

Two factors affect the precision of these measurements:

• The ratio of the timing period to the pulse period
• The accuracy of the timing

The first of these factors is a function of velocity. At high velocities, the pulse rate is high, so a large number of pulses are counted for each timing period. The precision is thus high also because small changes in velocity can be detected. At low velocity, however, the pulse rate is very low, so there will only be a few pulses in the timing period (or even none). The precision is then also very low since neighboring counts (e.g., 0 and 1) represent large (percentage) changes in velocity.

Increasing the timing period will increase the precision up to a point. But the problem requirements may dictate a maximum allowable sampling period (as in the previous velocity control example). The design solution is to increase the pulse rate by using a denser pulse-generating device, so at the minimum normal operating velocity, adequate precision will be obtained. This solution, though, can also have its problems, since at the maximum velocity the pulse-handling device will have to be able to process a higher pulse rate. The maximum rate limitation is of particular importance if the pulse counting is being done in software.

The timing accuracy is dependent on the frequency used to drive the timer and the latency between the end of the timing period and the moment at which the counter is read. It is also dependent on the latency between starting the timer and the counter, although that is not usually a problem. It is assumed that whether a hardware or software solution is used, a hardware timer will be utilized. Thus the timer's excitation frequency is the same in both cases. The time-out latency can differ greatly, however. In hardware solutions it should be possible to keep that latency period to within one tick of the timer's clock.

Software solutions, though, will use either interrupts, or poll the timer to find out when time is up. In the case of the interrupt, there is a fixed latency time associated with the interrupt setup and a variable latency associated with the specific instruction being interrupted. It is possible to compensate for the fixed latency in setting the timer, but the variable part of the latency will always cause noise in the counter reading. Polled solutions have completely variable latency, whose magnitude depends on the length of the polling loop.

10.2.6 Period Measurement

Mathematically, the period of the pulse signal carries the same information as the frequency. A division (inversion) is necessary to get the velocity from the period, which can be a problem in some implementations. The attractive feature of period measurement, though, is that its precision properties are exactly complementary to those of the frequency measurement methods. The period measurement is made by starting a timer at the time a pulse is detected, and stopping the timer when the next pulse is detected. Its precision is poor at high velocities when there are relatively few clock ticks between pulses, but the precision is good at low velocities when the pulses are widely spaced.

The main sources of error for period measurement are the same as for frequency measurement: latency between the detection of an event and the reading of a register. In the period measurement, the event detected is the arrival of a pulse and the timer's register is read; in frequency measurement, the event is the time-out and the counter's register is read. Another error source, uneven spacing of the pulse-generating mechanisms, tends to be more significant when period is measured. The vanes, slots, bands, and so on, that are used to produce the pulses can never be applied exactly evenly. Since the spacing between individual marks is measured, there will be a deterministic error in the velocity. Because the error is deterministic, it may be possible to compensate for it partially in some cases.

A variant of the period measurement, described in recent patents (U.S. 4,639,884 and 5,062,064), is to measure the period of a *group* of pulses. The group size is changed with velocity so that the amount of time for the group is roughly constant. This produces a precision that is more or less independent of velocity and, except at the lowest velocities, tends to average out the uneven spacing error. If period measurement is used in connection with a program that depends on equal sample-time scheduling, such as a digital control system, there will be an uncertainty in the sampling delay because the timing of the period measurement is entirely dependent on when the pulse occurs.

10.3 ANALOG POSITION MEASUREMENT

There are probably more ways to measure mechanical position than almost any other physical quantity. A few of these are described here. Those that are described are in common use and produce analog electrical signals, so they can be used with either analog or digital mechanical control systems. Variation in electrical resistance is one of the most common phenomena exploited for position measurement. It is very easy to implement, produces an analog signal directly, and is very low cost. The basic circuit is that of a potentiometer (See Figure 10.5 on next page). The wiper voltage, V_w, will depend on the ratio of resistances, as long as there is no current drawn through the wiper circuit,

$$V_W = \frac{R_{wg}V_{hi}}{R_{hg}}$$

(10.6)

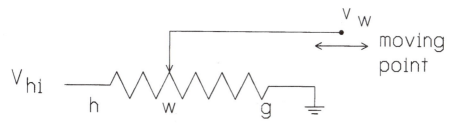

Figure 10.5 Resistance-Based Position Measurement

The no-current condition is easily approximated by using an amplifier with high input impedance, such as an operational amplifier, to isolate the wiper voltage from the downstream circuitry. The position measured is that of the part connected to the wiper relative to the part connected to the resistor. The calibration of the device depends on the geometric relationship of resistance to distance. It is usually made to be linear with distance (or linear with angle for rotary potentiometers), so that the output voltage is linear with distance (or angle).

For all its simplicity and low cost, however, resistance-based position measurement has serious limitations. Most of them are related to the mechanical interaction of the wiper and the resistance. The resistance can be made from a variety of materials, including wound wire and carbon. One problem is that the sliding contact generates electrical noise, which limits the precision and resolution of the sensor. Another is that the sliding contact wears, changing both the noise and the sensitivity of the instrument. Yet another is that the contact area is very susceptible to accumulations of dirt and contaminants, changes in humidity and temperature, and so on.

Noncontacting instruments avoid many of these problems. Two common noncontacting position-measuring instruments are based on variations in electrical capacitance and mutual inductance. Capacitance distance gages are generally useful for very small distances but have good linearity and precision within that range. Linear variable differential transformers (LVDTs) use a moving iron core and fixed coils. As the moving core is displaced, the mutual inductance between the fixed coils changes. If an ac excitation is used, these coils form a transformer with varying ratio, so the output (ac) voltage will change with position (Figure 10.6). To make an effective analog instrument, the ac voltage must then be converted to a dc voltage for output. LVDTs fall into the group of instruments that have all their electrical connections on one side of the motion, unlike the resistance/wiper instrument above. This is very useful when the mechanical parts are not physically connected in a way that allows for easy completion of the circuit.

Strain gages can also be used as position measurements. Strain gages use the resistance change as fine wires are stretched as measures of the internal strain in the part to which the gage is attached. They can be either ac or dc excited. Since they depend on the bending of a structural part, they can only be used in situations in which such parts are convenient for gage attachment. They are typically used for small displacements, for example, to measure the deflection of the load-carrying member in a force transducer.

Detection of magnetic fields with a Hall effect sensor can also be used as an effective position sensor. Hall effect sensors detect magnetic fields, so an arrangement of a permanent magnet and a Hall effect sensor can create a position sensor.

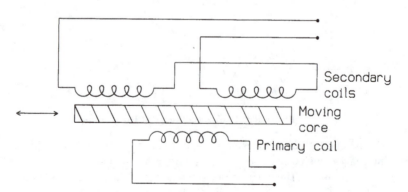

Figure 10.6 Linear Variable Differential Transformer

The magnetic field changes the pattern of current flow in a thin semiconductor sheet, thereby changing voltage differences across the sheet. The voltage changes are very small, 100 μV or less, so require amplification before a useful output can be produced. For example, if a system is built with the magnet moving and the part with the sensor stationary, the shape of the magnetic field will change as the relative positions change (Figure 10.7). This change in field can be detected by the Hall effect sensor to give a signal proportional to position. Again, this provides a noncontacting sensor with all the required wiring on one side of the moving boundary. Hall effect sensors can also be used as the presence detectors for pulse-generating velocity measurement systems. A disadvantage of Hall effect measurement is the need for the permanent magnet, which might cause attractive forces to be exerted that could themselves cause significant deflections.

10.4 INCREMENTAL ENCODERS

A real problem with the position sensors mentioned thus far is that it is difficult or impossible to use any of them to measure arbitrarily long distances as is common, for example, in machine tools. In most analog instruments, the physical configuration is not useful for long measurements. The potentiometer/resistance method can be used to produce instruments

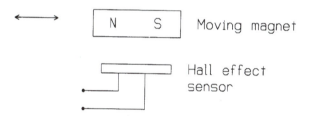

Figure 10.7 Hall Effect Position Sensor

with long ranges, but the relatively high noise level means that the measurement will suffer at the fine-positioning end if the length is extended. Encoders are inherently digital devices, so avoid the coupling of least-count resolution to maximum range. They are an extension of the pulse-based measurement systems for velocity, described previously, and have no theoretical limit on how long a measurement can be made.

10.4.1 Quadrature

A quadrature signal consists of two square pulse signals, 90° out of phase. Square pulse signals have 50% duty cycle, on half the time and off half the time (Figure 10.8). The important property of a quadrature signal is that both increments of position change and the direction of the change can be determined. The position is incremented at every transition, up or down, of the quadrature signal. As long as motion continues in one direction, the transitions alternate from *a* to *b* and back. Consequently, one channel will lag the other by 90°. On a change of direction, there will be two successive transitions on the same channel. The logic circuits discussed earlier can be used to interpret the quadrature signal, the decoding can be done in software, or specialized devices are commercially available to convert the quadrature signal to a digital position signal.

The first step in decoding the quadrature is usually to produce a pair of pulse trains, one for forward motion, one for reverse. These pulse trains can be used to drive up/down counters to give position information. The resolution of these pulse trains can vary depending on the decode logic used. Figure 10.9 shows typical 1X, 2X, and 4X decoding. 4X decoding decodes all edges of both channels a and b . The 1X, the lowest resolution, decodes only the positive edges on channel a. Direction is determined by examining the phase of the signal. In forward motion, as defined in the figure, positive edges on channel a are followed by positive edges on channel b. For reverse motion, positive edges on channel a are followed by negative edges on channel b. The 2X coding is intermediate. It decodes all edges of channel a, determining direction only by checking the sequence of positive to negative edges on channels a and b.

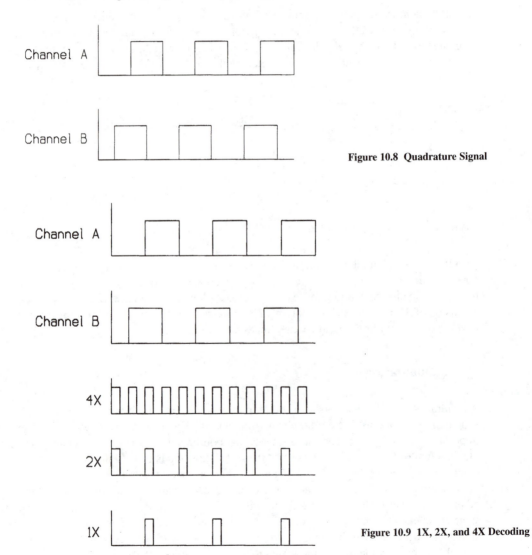

Figure 10.8 Quadrature Signal

Figure 10.9 1X, 2X, and 4X Decoding

10.4.2 Reliability

Incremental encoders, as the name implies, give only relative position information. An important disadvantage is that there is no information in the signal about absolute position. On power-up, for example, there is no way to know where the object is. It also means that the devices that decode the quadrature must be extremely reliable and not miscount *any* transitions. If they do, there is no way to check for the error, and such errors will accumulate. Because they are digital devices, however, it *is* possible to attain reliabilities that can count billions of transitions with no errors.

There is also information inherent in the signal that can be used to catch some errors. Because the quadrature signal is a Gray code, that is, only one bit changes at a time, any two-bit changes must be erroneous. Spurious pulses on a single channel, if they happen while there is no activity on either channel, will cause a count in one direction followed by a count in the other, for no net change, although the control system might react temporarily in response to the spurious pulse, causing some unnecessary vibration.

The nature of the electrical connections can make a large difference in the noise level in the signal. Use of differential signals can reduce noise levels greatly, especially where long wires are used or the environment is electrically noisy. Most systems that utilize incremental encoders use auxiliary signals to determine reference positions. Limit switches are commonly used, and most encoders are available with an additional channel to give a fiducial or index signal: for example, a pulse once per revolution. A common "homing" technique is to move the object until a limit switch is tripped. This gives a crude measure of the home position. The direction of motion is then reversed and a very slow movement is made until the encoder's index pulse is detected. This gives the finer measure of home.

10.4.3 Decoding Technologies

The most important characteristic of an encoder signal is the *rate* at which transitions occur. The maximum rate is determined by a combination of the maximum speed at which the object will move and the finest resolution needed for positioning. Maximum rates between 100 kHz and 1 MHz are not unusual. This rate determines the type of processing that must be used and to some extent the amount of noise to expect, since higher frequencies are more susceptible to noise. Software decoding can be used for rates up to about 1 kHz. An advantage to software decoding is that any necessary precision can be maintained in the count, with minor computing-time cost. Thus if it is necessary to keep a 32-bit count, or even higher, that can be done.

Higher rates require the use of external decoding devices. They are usually built from logic hardware and have a fixed counter width. Transferring the data from the decode device to the computer presents certain problems. First, if the count is kept in natural binary, which it usually is, multibit changes are possible. The transfer to the computer must be protected in some way against reading the data while they are changing, since a spurious result could be obtained if it did. A common method is to provide a transfer register. On command from the computer, the counter contents are moved into the transfer register. The computer can then read that register, secure that it will not change while being read.

The quadrature signal, however, is asynchronous to anything going on in the decoder to the computer. It will change at a random time with respect, for example, to the transfer of the counter contents to the register. The decoder must be designed so that regardless of the relative timing of events, no counts will be missed. When using such a device, either custom design or commercial, the computer must sample the position often enough to avoid ambiguity in the position due to rollover of the counter. Two methods of counting are commonly used. In one case, the counter is zeroed when its contents are transferred. Each time the transfer register is read, it thus gives the distance moved since the last time it was read. This is added to the current position in the computer, using any desired word size.

The alternative method is to allow the counter to free run. In that case the change in position is obtained by subtracting the previous reading from the current reading. Because the counter can go up or down, the count can be positive or negative, and since the counter is free running, can include a rollover (crossing of zero) in the counter. Fortunately, as long as the sampling is often enough, all of these complexities, including rollover, can be handled without difficulty. The count is in the form of a signed, 2's complement number. The subtraction operation works out properly for motion in either direction, with or without rollover, if the computer arithmetic is also done following the rules of 2's-complement arithmetic. For example, assume for illustration that the counter is three bits wide. At some moment the counter might read 101. If by the next sampling instant the system has moved ahead three counts, the counter reads 000 (counting up from 101: 110, 111, 000). Subtraction gives

$$000 - 101 = 011, \text{ borrow } 1$$

In two's-complement arithmetic, borrows and carries out of the highest place are ignored, so the result is that $011 = +3$ steps have been taken. The rollover past zero did not affect the result.

A negative motion of two steps would result in the next sample at 011, giving a motion of

$$011 - 101 = 110, \text{ borrow } 1$$

Again ignoring the borrow, $110 = -2$ steps. The counter itself is always treated as an unsigned number, even though the result of the subtraction is a two's-complement signed number. Thus, when doing mixed word-width computations, the counter is first extended as an unsigned number, (i.e, with zero fill to the left) up to the width of the computer word. The previous counter value is then subtracted from the current value, yielding the signed number representing relative motion.

The key to the sampling is that the counter must be sampled before an ambiguity arises as to which direction the count was in. For the three-bit counter, there must be no more than three forward or four backward pulses before sampling. For example, starting again with 101, if four forward steps are taken without sampling, the result will be 001. Subtraction gives

$$001 - 101 = 100, \text{ borrow } 1$$

In two's complement, this is a -4, which is completely wrong.

10.4.4 Optical Shutter

In an optical encoder, there is a fixed grate and a moving grate, with a light source on one side and a light receiver on the other (Figure 10.10). This arrangement for the shutter mechanism is used so that the size of the light source and receiver can be independent of the grid. For very large grids (i.e., very low precision encoders) it is possible to make the light transceiver smaller than the grid dimension. When this is the case, there need only be a moving grate. As it moves by the light it will alternately cover, then uncover, the receiver. Two receivers can be used to get quadrature with this method. When fine grids are used, however, the transceiver elements will be larger than the grid. In that case the fixed/moving grates must be used. In the fixed/moving arrangement, each phase of the quadrature uses a separate fixed grid, aligned with the appropriate phase difference. Many encoders actually use more than two receivers. A common arrangement is to derive each channel from a pair of receivers differing in phase by 180°. This push–pull arrangement improves the signal quality.

10.4.5 Microdecoding

The signal coming directly from an encoder's sensors may not be a digital signal. In the applications discussed previously, before any of the decoding, the signal was converted to digital form, for example, through the use of comparator circuits and Schmidt triggers. If the signal is decoded directly from the encoder outputs, however, it is possible to perform the philosophical equivalent of microstepping in stepping motors—that is, microdecoding.

With the moving grate of an optical encoder aligned with the fixed grate, the maximum amount of light reaches the receiver. As it moves, more and more of the light is blocked. If the opaque and transparent sections are the same length, the ideal signal from the light receiver will be a triangle wave, (Figure 10.11). The ideal shape depends on the fixed and moving grates having no clearance between them and the light source being perfectly collimated. In actual encoders, because of clearance, imperfectly collimated light, and other factors, the signal is more nearly sinusoidal. The actual light sources and receivers are usually larger than the size of a single grating, so that they average the light through several slits.

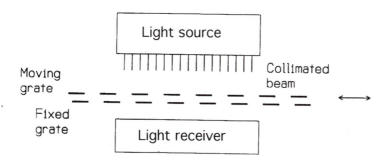

Figure 10.10 Encoder Light Shutter

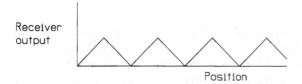

Figure 10.11 Ideal Receiver Output

In any case, as long as the shape of the signal is known, it can be fed to an analog-to-digital converter and through a table lookup to determine when a unit of motion has occurred. At that point, a quadrature transition on the digital output of the device is made. As with the purely digital quadrature, phases a and b are both needed to tell the direction of motion. Otherwise there is no way to know which side of the "hill" is being read at any moment. The number of microdivisions used depends on the repeatability of the analog signal, the noise level in the circuits, and the accuracy of the analog circuit elements.

10.4.6 Velocity from Encoders

Velocity can be derived from an encoder signal at several levels. The quadrature itself can be used, the forward/reverse pulse trains can be used, or the counter output. When using the quadrature signal or the pulse trains, the same methods can be used as were discussed in the velocity measurement from pulses, above. The added complexity is that there can be direction changes. At the moment a direction change is detected, the velocity is reported as zero. In both of these cases, either frequency or period measurements can be made. When the counter output is used to derive velocity, only frequency can be used. If the counter is sampled at regular intervals, the difference between the current and past readings is proportional to velocity. To improve the freshness of the velocity measure, the counter can be sampled more often than is needed for the control calculations. By using the most recent value, the value of velocity used for control will be closer to the instantaneous velocity. If the counter is sampled only once, the velocity used is the average of the velocity over the entire sampling period. This method of frequent sampling for velocity can be used as long as the velocity is high enough. At low velocity, the precision of the velocity measurement will become too low, and the full sample time will have to be used.

10.4.7 Linear and Rotary Encoders

Encoders can be built for either linear or rotary motion. Because the output is incremental in either case, there is no fundamental difference in their use. Rotary motion can be measured directly by a rotary (shaft) encoder. Linear motion provides three choices for configuration. A linear encoder can be used, a rotary encoder can be attached to a pinion gear riding on a linear rack, or a rotary encoder can be attached to the (rotary) motor drive shaft. As a generalization, the linear encoder would be used in applications requiring the greatest accuracy because it is coupled directly to the target motion. Use of a linear rack and rotary encoder is an intermediate solution. There are inaccuracies associated with the gearing, but there is no load on it other than the parasitic load of the encoder, so the deflections are minimized and antibacklash gears can be used. In some instances, the linear rack/rotary encoder solution may be more economical than a linear encoder. Measuring from the motor shaft is the least accurate but most economical solution. Because the drive train between the motor shaft and the load carries the full drive power, there could be significant differences between shaft position and load position, due to deflection of mechanical components.

10.5 LASER INTERFEROMETRY

Laser interferometers are in a class by themselves, although it would also be correct to group them as a type of incremental encoder. What sets them apart is precision, 10 to 100 times as precise as an optical incremental encoder, and cost. Laser interferometers measure distance within a wavelength of light by examining the phase relationship between a reference beam and a light beam reflected from the target object. The role of the laser in these systems is to produce light having only a single frequency present with phase coherence. Interferometry is impractical without such a light source because the mixed frequency/phase light of normal sources does not have a well-defined phase to compare.

The output depends on mixing of two sinusoidal signals, of the same frequency but different phase,

$$sin(wt) + sin(wt + \phi) \tag{10.7}$$

yielding a modulated sinusoidal signal with its peak amplitude occurring when the two signals are in phase, and minimum amplitude (zero if the amplitudes are exactly matched) when the two signals are 180° out of phase. The physical arrangement to achieve this is shown in Figure 10.12. The laser beam is passed through a collimating lens to make the

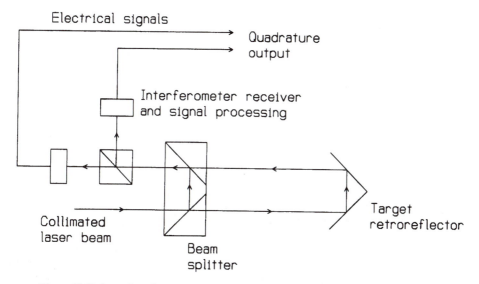

Figure 10.12 Laser Interferometer

rays as parallel as possible, then through a beam splitter. The beam splitter sends part of the light energy to the target, and the rest becomes the reference signal. The beam to the target is sent back by a retroreflecting prism or a mirror. The reflected beam is rejoined with the reference beam in the other half of the beam splitter and then sent to the detectors, which demodulate the signals to produce signals proportional to the amplitude of the incident signal.

The phase difference between the reflected and reference signals is equal to the remainder of the path length divided by the wavelength of the light, (i.e., the fraction of a wavelength left over when all the integral wavelengths are removed). There is no way to know how many whole wavelengths there are in the path, so the laser interferometer is inherently an incremental instrument. The total path length is twice the distance from the beamsplitter to the target, so there will be a full wavelength change in phase when the target moves by one-half the wavelength. The two detectors are arranged so that they differ in phase by one-fourth of a wavelength. The signals they produce thus represent phase differences from their inputs of 90°, so are sinusoids in quadrature, similar to the signals produced by the shutters of an optical incremental encoder.

If these signals are digitized immediately, a digital quadrature signal results which can give up to four transitions per complete cycle. This gives a resolution of one-eighth of the wavelength of the source light, since a target motion of one-half the wavelength generates a full cycle on the output. For a helium–neon laser, with a wavelength of about 0.6 meter (25μ in.), the base resolution is about 0.08 m (3μ in.). If the output signals are processed before digitization, resolutions as low as 5 nm (0.2 μin.) are possible. The major sources of errors in laser interferometry come from mechanical misalignments in the instrument mounting and from changes in temperature, which affect both the mountings and the optical properties of the light path. The temperature effects, in particular, can be compensated since the optical properties as a function of temperature are well known.

10.6 SYNCHROS AND RESOLVERS

These devices often compete with incremental encoders for motion control applications. In many ways, synchro/resolvers and incremental encoders are complementary methods for motion measurement. Synchros and resolvers are analog devices that measure angle absolutely and, with proper signal processing, can also produce a velocity signal, whereas encoders are digital, relative, and velocity can only be derived indirectly. Synchros and resolvers are constructed with materials and components similar to those used to make motors, so tend to have the same environmental constraints as those of the motors to which they are attached.

10.6.1 Inductive Coupling

Synchros and resolvers use variations in inductive coupling to measure angular displacement. Unlike the LVDT, which utilizes a moving core to change the coupling between coils, the coils themselves are moved in synchros and resolvers. A synchro or resolver has a set of rotor coils and a set of stator coils. The connection to the rotor coils is brought out through slip rings with brush contacts. In contrast to the brush connections in motors, however, the slip rings are continuous rather than segmented since no commutation is required. Continuous slip rings do not cause as much noise, or wear as much, as segmented commutators.

Ac excitation is used, usually applied to the rotor coil(s). The combination of rotor coils and stator coils forms a transformer, but it is a transformer with variable coupling because the rotor coil moves. Both the resolver (See Figure 10.13 on next page) and the synchro (See Figure 10.14 on next page) use single-phase ac excitation on the rotor coil. The resolver has two sets of isolated coils on the stator, giving it a four-wire isolated pair output. The synchro has a Y-connected set of three coils on the stator, so it has a three-wire output.

10.6.2 Input–Output Relationships

The outputs are ac signals at the same frequency as the excitation, w . The *amplitudes* of these signals are proportional to sinusoids of the shaft angle, o , because the relative coupling of the coils depends on shaft position. The ideal resolver outputs are

$$V_x = (K \sin \theta) \sin\omega t \tag{10.8}$$

$$V_y = (K \cos \theta) \sin\omega t \tag{10.9}$$

The parentheses have been added to emphasize the variable magnitude of the ac *carrier*. At constant shaft angle, the two signals are in-phase sinusoids with the same frequency but different amplitudes. It is the difference in amplitude from which the angular position of the shaft will be computed.

For an ideal synchro, the output voltages are

$$V_{1\text{-}3} = (K \sin \theta) \sin\omega t \tag{10.10}$$

$$V_{2\text{-}3} = [(K \sin(\theta + 120)] \sin\omega t \tag{10.11}$$

$$V_{2\text{-}1} = [(K \sin(\theta - 120)] \sin\omega t \tag{10.12}$$

Again, the amplitude differences among the signals contain the relevant information from which shaft angle can be deduced. Although synchro and resolver processing is often thought of as a phase-detection process, it really isn't. At least ideally, all the signals have the same phase and it is amplitude detection (demodulation) that provides the desired position information.

The equations above are the ideal relationships. In actuality, there are several factors that cause deviations from this ideal. Two of the most important are phase shifts between the input excitation and the output signals and the voltage induced by rotor motion, which is equivalent to back-emf in a motor. These and other deviations from ideal behavior have to be handled by the signal-processing components used to derive the position and, may, in some cases, establish operating limits for an instrument.

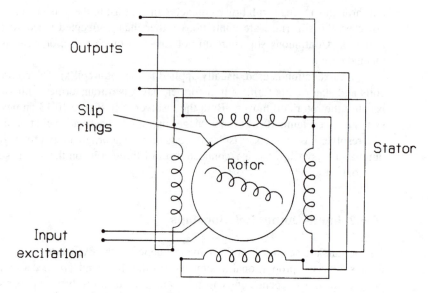

Figure 10.13 Resolver Schematic

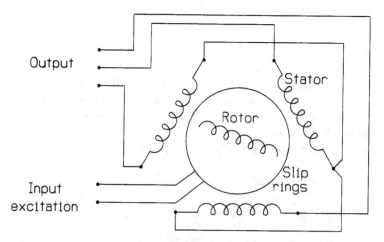

Figure 10.14 Synchro Schematic

10.6.3 Conversion to Position and Velocity Signals

Figure 10.15 shows an idealized resolver to the position–velocity conversion system. It is idealized in the sense that all sources of noise and error are ignored, and also because no indication is made of how the computational functions are implemented. It does, however, follow the logical structure of a class of commercial resolver-to-digital converters. The resolver input format is used because it is computationally simpler and is much more commonly used. Even if a synchro is used, however, its output can be converted to resolver format with a special-purpose transformer.

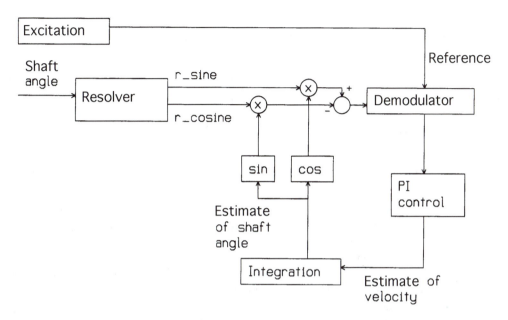

Figure 10.15 Resolver Signal Converter

The resolver has two inputs, excitation and shaft angle. The excitation is a single-phase ac signal. The shaft angle is the process variable being measured. The resolver produces two outputs, r_sine and r_cosine, which are described by equations (10.8) and (10.9). These signals are shown in Figure 10.16 for one complete revolution of the shaft. The top graph shows the shaft angle going from 0 to 360° over the time period indicated by the scale. The two lower graphs are the resolver outputs. The dominant frequency is the excitation or carrier frequency. The motion of the shaft modulates the amplitudes of the two signals according to the sine–cosine relationships. The task of the conversion system is to deduce the shaft angle by extracting that amplitude information from the resolver outputs.

The conversion process uses a computational feedback loop to compare an estimate of the shaft angle (ϕ) to the actual shaft angle (θ), and correct the estimate according to the error between them. Because of the nature of the resolver signals, the comparison is done by utilizing trigonometric relationships. The first step is to take the sine and cosine of the estimated shaft angle. The sin θ signal is multiplied by the cos θ signal from the resolver, and vice versa. This gives the two signals

$$\cos \phi \sin \theta \sin \omega t \qquad (10.13)$$

$$\sin \phi \cos \theta \sin \omega t \tag{10.14}$$

The coefficient, K, has been set to unity for convenience. When these are subtracted, the trigonometric relationship

$$\cos \phi \sin \theta - \sin \phi \cos \theta = \sin(\theta - \phi) \tag{10.15}$$

is used to show that the resulting, still ac, signal is

$$\sin(\theta - \phi) \sin \omega t \tag{10.16}$$

The interesting thing about this signal is that it contains in it the difference, $(\theta - \phi)$, which is the error between the estimated shaft position and the actual shaft position. Since the signal is still an AC signal, though, that property cannot yet be utilized.

The next step, then, is to remove the ac component by running the signal through a demodulator, as shown. In practice, much of the error rejection tasks fall to the demodulator. In this idealized example, though, the ac component will be removed by dividing the demodulator's input signal by the excitation (reference) signal itself. This signal is produced by the instrument, so is known and available. The output of the demodulator is thus

$$\text{demod} = \frac{\sin(\theta - \phi)\sin \omega t}{\sin \omega t} = \sin(\theta - \phi) \tag{10.17}$$

This is easily computed, except when the reference signal is zero. For that reason the computation is applied only when the absolute value of the reference is above a specified limit. Otherwise, the most recent output value is used. This is a dc signal now, with a magnitude corresponding to the sine of the error in angle. For small errors, it is, in fact, equal to the error. This signal is used as the input to the two-stage control computation, which has ϕ_{est} as its output. The first stage is a proportional plus integral (PI) computation, and the second stage is pure integrator. The two stage-integration is used so that there will not be any angle error during times when there is a constant angular velocity. When there is a nonzero acceleration, there will be a transient error in angle, which will persist until the acceleration becomes zero.

Use of a double integration thus achieves better performance from the instrument with respect to errors during constant-velocity (and low-acceleration conditions), but such a scheme is harder to tune and does not have as good transient behavior as a controller with only a single integration. The constant or near-constant-velocity situation is so important, however, that commercial converters usually use double integration. This closes the loop; the estimated angle is produced by the second integrator and is fed back

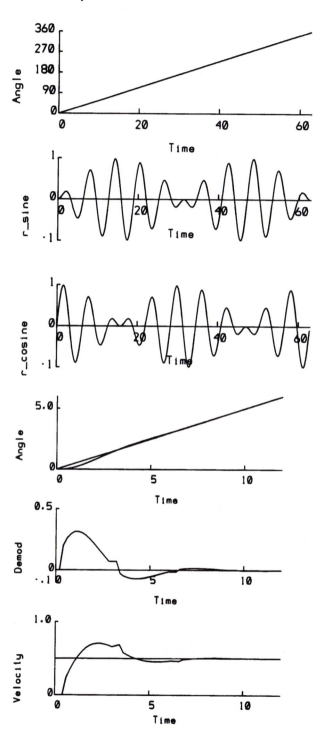

Figure 10.16 Resolver Output Signals

Figure 10.17 Resolver Signal
Conversion Output

to the multipliers. Figure 10.17 shows the results of a simulation experiment using this method. The top of the figure shows angle, actual and estimated. The system was initially at rest at its zero position, so the sudden change in angular velocity represents a transient condition. (The time scale and angular velocity are not the same as in the earlier figure.) The estimated angle is seen to converge to the correct value very quickly and to remain with no error. The graph labeled "Demod" is the output of the demodulator, which is an error signal. It shows an increase to a large transient error, which then dies away to zero. The action of the demodulator when the reference signal is small is also shown on the graph. This is the small section of the curve during which the error is constant (horizontal).

10.6.4 Velocity

The resolver conversion also has the very useful property that a velocity signal is generated along with the angle signal. Referring again to Figure 10.15, the block diagram of the conversion system, there is a signal marked "estimated velocity." This is the input to the integrator that produces the estimated angle. Since the two signals differ by an integration, the input to the integrator must be the derivative of the output, or the rate of change of the estimated angle, that is, the estimated velocity. The bottom graph in Figure 10.17 shows the actual and estimated velocity. It also tracks with zero error once the transient has died away. It is interesting to note that the transient behavior of the velocity signal is not as good as that of the angle signal. It takes longer to reach the tracking velocity, deviates more, and is probably more sensitive to noise.

10.6.5 Resolver-to-Digital Converters

Tracking-type resolver-to-digital converters follow the same basic computing scheme as that outlined above. The major structural difference is that the second integrator is replaced with an up/down counter, and the output of the PI block is input to a voltage-controlled oscillator (VCO), which produces the pulses to drive the counter. This provides the same computational function as outlined previously. The demodulator is a key component in real systems. It is called on to perform much of the noise rejection in the instrument. In particular, it must reject out-of-phase signals such as the speed velocity (back-emf) and correct for small phase errors in the resolver signals.

   The counter output is the digital output of the converter. The velocity signal remains in analog form, where it can be used in place of a tachometer signal. It can be converted to digital form with an analog-to-digital converter. The demodulator can be avoided by using a successive-approximation sampling converter, but the overall scheme remains very similar. Overall precision of 12 bits (1 part in 4096) to 14 bits (1 part in 16,384) or more is readily attainable with resolver-to-digital converters. Higher precision is traded against update rate in sampled-type converters.

10.6.6 Rotary/Linear Conversion, Multistage

Synchros and resolvers are available only as rotary devices. They must be used upstream of the rotary/linear conversion in a device that ultimately produces linear motion or must be attached to a separate rack and pinion, as described previously for rotary encoders. To retain their absolute properties, however, they must be geared in such a way that the full range of motion falls within a single revolution of the synchro or resolver. In practice, many, if not most, motor-driven systems rely on multiple revolutions of the motor to achieve their full range of motion, so special gearing must be used to attach the synchro or resolver.

This form of gearing can also be used to increase the dynamic range of synchro/resolver systems by using multistage devices. Because they are analog, the dynamic range is inherent to the instrument. It can be improved by better, more expensive design, but there is always a limit. That limit can be extended by using two synchros, one for coarse motion and one for fine motion. This can be accomplished by having a specified gear ratio separating the two stages. The lowest reliable bit of the coarse instrument's output establishes a datum for the fine instrument. By coordinating the gear ratios with this bit value, the fine instrument will measure position within that range. The method works because of the absolute nature of synchros and resolvers. There is never any question as to the relative locations of the coarse and fine instruments, even at power-up.

10.7 ABSOLUTE ENCODERS

An absolute encoder is a multichannel extension of the incremental encoder. It is used in situations where an absolute position is needed in digital form, and in that regard, is an alternative to use of a resolver. Absolute encoders use one track for each bit of resolution in the position signal. The output is an n-bit parallel signal giving an absolute measure of the position within a single revolution of the encoder (absolute encoders are normally available for rotary motion measurement). The signals are most commonly generated optically or using brushes to establish an electrical conduction path.

Each track is a concentric ring on the code disk with alternating areas of clear/opaque, reflecting/nonreflecting, or conducting/insulating, depending on the technology used. There are three common codes that are generated; the code can be determined by the patterns used on the code disk or by the electronics used to interpret the output signals. The codes are:

- Natural binary
- Binary-coded decimal (BCD)
- Gray code binary

Natural binary is the usual binary (unsigned) integer format. A three-bit code wheel implementing natural binary coding is shown in Figure 10.18. In this case, the outer ring is the least significant bit. If the wheel is rotating at constant speed, the signals derived from this encoder will be as shown in Figure 10.19, with the most significant bit at the top. The numerical sequence is

000
001
010
011
100
101
110
111

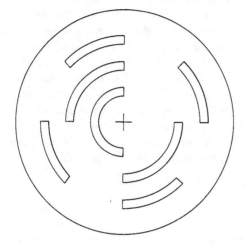

Figure 10.18 Natural Binary Code Disk

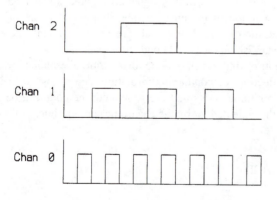

Chan 2

Chan 1

Chan 0

Figure 10.19 Signals for a Natural Binary Code Disk   ·

BCD (binary-coded decimal) uses a coding scheme similar to that of natural binary, except that the bits are grouped four at a time, with each four-bit grouping corresponding to a single *decimal* digit. Since a decimal digit uses only 10 of the possible 16 combinations, there is an inefficiency associated with using BCD. It is, however, convenient for interfacing with operator displays and some instruments, and can be used for internal arithmetic in some computers.

An eight-channel absolute encoder using BCD can code two decimal digits to give output values from 0 to 99. The same eight channels, encoded in natural binary, have outputs from 0 to 255. There is, however, an interfacing problem associated with natural binary (and BCD). In the change, for example, from position 3 (011) to position 4 (100), all three of the bits have to change. It is physically impossible for all three of them to change at exactly the same time. If the output signal is read while a change is in progress, it is possible that the value read will reflect changes in fewer than the three bits that need to change and thus will be a completely erroneous value. For instance, if the lowest-order bit changes first, it would be possible to read the value 010, which is 2, instead of the correct value of 100 (4).

There are two solutions to this problem. One is to provide additional logic to assure that the result is read only when it is stable. Since the motion of the code disk is asynchronous with the electronics, information must be encoded on the disk itself to indicate regions of validity. Use of time delays or other time-related schemes will not work because the disk can be turning at arbitrarily low speeds, so there is no way to know if the transition is complete by waiting. The other solution is to use codes on the disk that do not have the multibit-change problem.

The design of codes that have only one bit changing between adjacent values was discussed in Chapter 2. *Gray codes*, which were discussed in the context of construction of logic design (Karnaugh) maps, have that property and are widely used for absolute encoders. A three-bit Gray code is

000
001
011
010
110
111
101
100

The code wheel for this Gray code and its associated signals are shown in Figures 10.20 and 10.21. Use of a Gray code disk removes the potential for error due to multibit changes. Even if a transition is under way, the only possible values that can be read are the *before* and *after* positions, either of which is a valid result.

Most optical absolute encoders are made with Gray-coded disks. The natural binary or BCD options are produced electronically. For natural binary or BCD, the very fast transition speed of the electronics is relied on to give reliable readings. If the Gray code option is used, a code conversion must be made. For low-precision systems, a table-lookup or static logic circuit can be used. As the precision increases and more channels are used, the number of possible combinations gets very large, so this option is not feasible. A simple sequential algorithm can be used in its place, which is applicable to either hardware or software implementation.

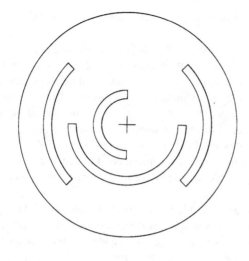

Figure 10.20 Gray Code Disk

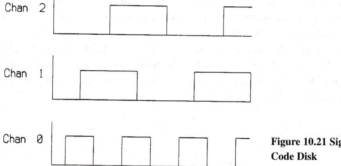

Figure 10.21 Signals for a Gray Binary Code Disk

Consider a binary signal (natural or Gray) consisting of n bits numbered from ($n-1$) to 0, with the low-order bit numbered 0. The natural binary (b's) equivalent to a Gray binary (g's) signal is

$$b_{k-1} = b_k \underline{V} g_{k-1} \tag{10.19}$$

where the $\underline{V}$ symbol is for the exclusive-OR operation. The computation starts with k=n. The initial value of b_n is set as 0 (this value is used only to start the computation; the actual result only includes up to b_{n-1}). This can be implemented in hardware with shift registers and gates, or can be programmed directly for software implementation.

10.8 PROBLEMS AND DISCUSSION TOPICS

1. Experimentally determine the dynamic range of a velocity measurement based on an analog tachometer driven by a dc motor. Use the highest-precision A/D converter available and take data over a wide range of velocities. If the noise level at the low end of the velocity range falls below the precision available, amplify the signal from the tachometer for use at low velocity to attempt to determine the noise level. Make sure to determine the noise level of the A/D converter itself (by grounding its inputs)!

2. Equations (10.4) and (10.5) give the continuous-time representations for a lead–lag element. A

simple difference approximation,

$$\frac{dx}{dt} = \frac{\Delta x}{\Delta t} \tag{10.19}$$

can be used to convert it to a discrete-time form suitable for computer use.

(a) Implement the lead–lag on a computer with A/D and D/A and verify the basic step responses shown in Figure 10.1. The easiest way to do this is to use a standard signal generator set to output square waves. Adjust the frequency so that the lead–lag output response has decayed almost completely before the next square-wave edge occurs. Explore the range of parameters to see the responses obtained.

(b) Connect the lead–lag controller to the motor of Problem 1 and find the compensation parameters giving the best response. Try it with both voltage and current amplifiers.

(c) Add inertia to the motor and retune the compensator, observing how the parameters change with added inertia.

(d) Add a frictional load to the motor and, again, retune the controller.

3. Explore the dynamic range for as many analog position sensors as can be assembled.

4. As the velocity becomes small, noise becomes a major factor in the quality of control that can be achieved. The simplest noise filter that can be implemented on a computer is boxcar averaging (i.e., averaging the most recent N samples). On the other hand, the added dynamics introduced by the filter reduces the maximum controller gains that can be used, so reduces control quality.

(a) Try implementing a velocity filter at fairly high velocity (where noise should not be a significant problem) and see how much deterioration there is in control quality (if any).

(b) Now try the same filter at low enough velocity that noise is a problem and see if there is any improvement in control quality (a motor with good mechanics is useful for this work; otherwise, the low-velocity behavior may be dominated by stiction).

5. Implement a position control using a cascade control with analog velocity and position sensors. Explore point-to-point moves by making the same move many times and recording the actual stopping positions.

6. Design and build both synchronous and asynchronous circuits for decoding quadrature from an encoder. Decode to forward/reverse pulse trains and feed these into an up/down counter. Use an analog position sensor to verify the decoding results.

7. To avoid problems with reading a natural binary counter into a computer (reading while bits are changing), design a quadrature decoder that produces a Gray code result. Devise a test that can detect if an erroneous result was read in. Using that test, see if the error rate due to asynchronous reads can be measured for natural binary and Gray code decoders.

8. As another defense against asynchronous read errors, design a synchronous circuit that can transfer a result from the counting register to a read register on command from the computer. The clock rate of the synchronous circuit sets a maximum quadrature rate that can be handled. Determine what that rate is for the circuit you designed.

9. Try measuring velocity from a motor equipped with an encoder by "time stamping" each encoder transition. Some microcontrollers have such circuitry built in. If not, it can be done (with somewhat less precision) by attaching an interrupt to the encoder input and recording the times at which the interrupts occur. This will work only at low velocities but can greatly improve the quality of the velocity measurement.

10. There are special-purpose chips for decoding quadrature (from Hewlett-Packard, for example).

Connect one to the digital input port of a microcontroller or computer and use it to implement a feedback position control. Run the quadrature signals through a buffer, then implement the velocity measurement scheme of Problem 9 in parallel with a decoder chip.

11. Resolvers are usually used in conjunction with resolver-to-digital (RTD) specialized circuits. Interface an RTD to a computer and use it to implement a cascade position control. Compare the quality of the RTD velocity signal to the signal obtained from an analog tachometer.

11

Operational Amplifiers for Analog Signal Processing

Operational amplifiers can achieve nearly effortlessly what is the subject for extensive research in digital systems: parallel computation. Operational amplifiers are also known as computing amplifiers because of their ability to realize easily such common computing functions as summation, integration, and differentiation. Complex functions can be realized by combining elements in parallel or series. The analog computer is based on operational amplifiers, and although supplanted by digital computers for many general-purpose applications, retains an important role in simulations that require rapid, real-time response, such as trainers.

Op-amps, as operational amplifiers are often called, derive much of their popularity from the property that, in most applications, the characteristics of the final circuit depend only on the properties of the passive components (resistors, capacitors, etc.) and not on the properties of the amplifier. Highly accurate and stable passive components are much easier and cheaper to obtain than are highly accurate active components. Despite increased "digitization" of mechanical systems control, operational amplifiers remain critical components for mechanical system interfacing. They are the standard building blocks for analog signal processing at medium power levels, that is, once the output from low-powered transducers have been through first-stage amplification and prior to the power amplifier in the actuation stage. They are used for such applications as isolation, filtering, nonlinear functions such as rectification, peak detectors, comparators, and voltage–current converters.

11.1 HIGH-GAIN DC AMPLIFIER

An operational amplifier is a very high gain dc amplifier with differential inputs and, usually, single-ended output. It is most normally operated with a bipolar, balanced power supply, giving an output range that spans positive and negative voltages. The circuit symbol for an operational amplifier is shown in Figure 11.1. It shows only the signal lines; the power supply connections are not shown. Internally, an operational amplifier consists of a multistage transistor amplifier. The popular 741 operational amplifier, for example, contains 20 transistors plus resistors and capacitors. As shown in the figure, the operational amplifier is not very useful. It has a gain of 10^5 or greater, so that an input voltage difference of about 0.1 mV will cause the amplifier to reach its maximum output. This very high gain, though, is not a reliable parameter in the sense that the actual value of the gain can be counted on. It is thus not useful as a high-gain instrumentation amplifier.

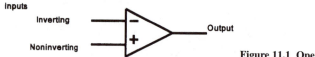

Inputs

Inverting

Noninverting

Output

Figure 11.1 Operational Amplifier

Its usefulness comes from adding feedback elements to the circuit in such a way as to degrade the gain. This seemingly anomalous behavior is actually the trick that makes the operational amplifier what it is. It is one of those inventions that seems perfectly obvious in hindsight. The operating characteristics of simple operational amplifier circuits (most applications of operational amplifiers require nothing more than simple circuits) can be deduced by starting from the characteristics of an ideal operational amplifier:

- Infinite open-loop gain
- Infinite input impedance (draws no current)
- Zero output impedance (voltage independent of load)

Figure 11.2 shows a circuit using the inverting input, with an input resistor and a feedback resistor. Application of these ideal characteristics leads to the important conclusion that V_- must be very close to zero. This is a consequence of the high gain. If the output is at a reasonable voltage, which it must be because it cannot exceed the supply voltage, the difference between the inputs must be very small. Since V_+ is at ground, V_- must be very near ground. The feedback and input current flows can now be calculated:

$$i_f = \frac{V_{out}}{R_f}$$

(11.1)

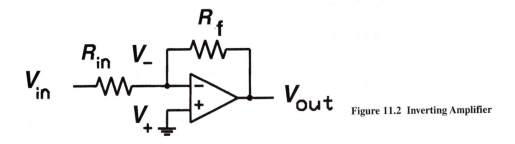

Figure 11.2 Inverting Amplifier

$$i_{in} = \frac{V_{in}}{R_{in}} \tag{11.2}$$

Since the amplifier ideally draws no current at its inputs, these currents must be equal and opposite,

$$i_f = -i_{in} \tag{11.3}$$

Therefore, substituting and solving yields

$$V_{out} = -\frac{R_f}{R_{in}} V_{in} \tag{11.4}$$

This application shows the typical characteristics of operational amplifier systems. The feedback resistance provides negative feedback that serves to reduce the overall gain. Because the current at the inverting input, which is also referred to as the summing junction, must sum to zero, any small change in the output voltage will cause a voltage imbalance at the summing junction. Because of the high open-loop gain of the amplifier, that would cause a very large change in the output voltage to drive the current in the other direction.

The overall properties of the circuit are indeed independent of the characteristics of the amplifier. The resistor values alone determine the circuit behavior, at least in the ideal case. Another common characteristic of computing-type circuits, often an annoying one, is that the output voltage is the negative of the input voltage. The circuit input–output impedance characteristics are determined on the input side by the size of the input resistor used, and on the output side by the amplifier's output impedance. Input resistors from 10 kΩ to 1 MΩ are common, so the input impedance is fairly low. Signals from high-impedance sources (i.e., instruments that are not capable of supplying much current) may require some amplification before being used with an amplifier of this sort. The output impedance is usually quite low and rarely a problem. When using an operational amplifier in this configuration, the inverting input is called a *virtual ground*. With the + (non-inverting) input grounded, the - (inverting) input will always have a voltage very close to ground. It is not *exactly* ground, however. The small voltage difference to ground (whose magnitude is determined by the amplifier's gain) is what makes the circuit function.

11.2 COMPUTING FUNCTIONS USING INVERTING AMPLIFIER

11.2.1 Static Linear Functions

The inverting amplifier implementing a gain, shown previously, can be extended to function as a summer. The principle of the summer is that the currents must sum to zero at the inverting input, the summing junction. In effect, any number of input currents must balance the current coming through the feedback resistance. For a circuit of the form of Figure 11.3, the current equation is

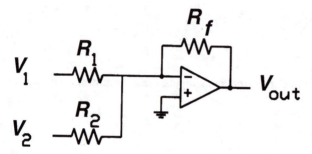

Figure 11.3 Summing Amplifier

$$i_f = -(i_1 + i_2) \qquad (11.5)$$

Since the summing junction remains at near-zero voltage, these currents are easily computed,

$$i_f = -\left(\frac{V_1}{R_1} + \frac{V_2}{R_2}\right) \qquad (11.6)$$

and the output voltage is

$$V_{out} = -\frac{R_f}{R_1} V_1 - \frac{R_f}{R_2} V_2 \qquad (11.7)$$

Any number of additional inputs can be summed in the same manner.

11.2.2 Dynamic Linear Functions

The gain and summer configurations used only resistors in the input and feedback paths. Resistors are *static* elements (i.e., they store no energy), so the resulting circuits have static characteristics. In a static system, the outputs depend entirely on the *current* values of the inputs.

  Dynamic functions, functions whose outputs also depend on the past history of the inputs, can be constructed by using energy storage elements in the feedback and/or input circuits as well as resistors. Capacitors are normally used because they are more widely available than inductors and much less expensive.

Figure 11.4 shows a circuit with a capacitor in the feedback path. The analysis follows the same pattern as used previously: equation (11.3), the current summation equation applies, as does equation (11.2) for the current in the input path. The feedback current can be computed from the constitutive equation for a capacitor,

$$i_f = C_f \frac{dV_{out}}{dt} \qquad\qquad (11.8)$$

To solve this for the output voltage, the differentiation must be inverted and expressed as integration,

$$V_{out} = -\frac{1}{R_{in} C_f} \int V_{in}\, dt \qquad\qquad (11.9)$$

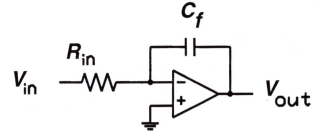

Figure 11.4 Integrating Amplifier

This circuit therefore behaves as an integrator (with the ubiquitous sign reversal), with the output voltage dependent on the integral of the input voltage, scaled according to the resistance and capacitance values. Notice, again, that the functional characteristics depend only on the values of the passive component properties. That this is a dynamic circuit is seen from the behavior of the integrator, which is dependent on the complete past history of the input voltage.

If the capacitor and resistor are switched, with the capacitor in the input path and the resistor in the feedback path (Figure 11.5), the output voltage will be

$$V_{out} = - R_f C_{in} \frac{dV_{in}}{dt} \qquad\qquad (11.10)$$

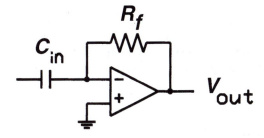

Figure 11.5 Differentiator

This circuit behaves as a differentiator, with the output equal to (the negative of) the derivative of the input. Differentiators must be used with caution, however, because they are very sensitive to noise.

There is no restriction to using single elements in the feedback or input paths. The circuit of Figure 11.6 uses a capacitor and resistor in parallel in the feedback. The feedback current for this circuit is

$$i_f = \frac{V_{out}}{R_f} + C_f \frac{dV_{out}}{dt} \tag{11.11}$$

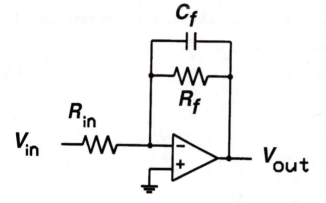

Figure 11.6 Computing Op-Amp Circuit for a Low-Pass Filter

Equating this to the negative of the input current yields a differential equation for the output voltage,

$$\frac{dV_{out}}{dt} + \frac{1}{R_f C_f} V_{out} = -\frac{1}{R_{in} C_f} V_{in} \tag{11.12}$$

which is equivalent to the transfer function,

$$V_{out} = \frac{-\dfrac{1}{R_{in} C_f}}{s + \dfrac{1}{R_f C_f}} V_{in} \tag{11.13}$$

This differential equation describes a first-order low-pass filter. The output signal will be similar to the input signal except that the higher-frequency components will be removed. A filter of this sort could be used, for example, as an antialiasing filter for an analog-to-digital converter. The scope of dynamic functions that can be realized in this manner is limited only by the imagination of the designer. With proper design, the input and feedback circuits will completely specify the circuit's behavior, as with the circuits examined thus far.

11.2.3 Analog Computers

Analog computers are used to solve differential equations by interconnecting summers, integrators, and possibly, nonlinear elements. The main component of an analog computer is an integrating amplifier with an additional circuit added so that the initial voltage across the capacitor can be controlled (Figure 11.7). At "time zero" the switch is opened and the integrators can free-run to produce the solution to the differential equation.

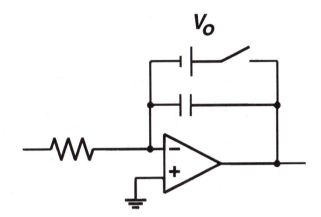

Figure 11.7 Analog Computer Integrator with Initial Condition

To prepare for analog computer solution, differential equations of order greater than 1 must be expressed as a set of coupled first-order equations,

$$\frac{dx_1}{dt} = f_1(x1,x2,...,t) \tag{11.14}$$

$$\frac{dx_2}{dt} = f_2(x1,x2,...,t) \tag{11.15}$$

and so on. The next step is to *scale* the equation. The analog computer uses voltages for all of its dependent variables, and time as its independent variable. The abstract variables from these equations must be replaced by new variables using the units available in the analog computer. Each of the dependent variables (x's) is replaced with a variable in units of volts, and the independent variable (t) is replaced by a new variable with units of real time. The scaling factors used must produce an analog computer circuit that uses reasonable component values (resistors and capacitors) and that does not exceed the allowable voltage range. Furthermore, the scaling must be done in a way that does not make any of the variables so small that they are overwhelmed by the noise in the circuit. Scaling is particularly difficult when nonlinear elements are used.

The process of scaling for analog computer solution is very similar to the scaling process needed to utilize integer computation on a digital computer. The scaled equation is then realized with a circuit. The mathematical complexity of the original system is manifested in the size of the resulting circuit. Unlike a digital solution, the computational speed is independent of the problem complexity. All of the elements operate simultaneously, so a parallel solution is realized. Because general nonlinear elements are difficult to build for analog systems, digital solutions are now almost universal for purely computational problems. Analog computers can still be useful in hybrid operations, where the analog computer is used in conjunction with a digital computer to utilize its parallelism and high computation speed, and in real-time simulators, used most often for training.

11.2.4 Nonlinear Functions

The lowly diode is the key to unlocking many of the nonlinear possibilities in operational amplifier circuits. An ideal diode will conduct current in only one direction. It thus acts as a zero resistance for forward voltage bias (in the direction of the arrow, Figure 11.8a) and an infinite resistance for reverse bias. Real diode behavior is more like the graph Figure 11.8b. There is a small but finite resistance for forward biasing, and more important, there is an offset voltage above zero that is required to enter the conduction state. This voltage depends on the diode's construction; silicon diodes have a forward voltage drop of about 0.6 V. In the reverse direction, the resistance is very large, but not infinite. Reverse currents of modest fractions of a microampere are usual. The reverse-biased diode thus acts as close to an open circuit as long as the applied voltage does not exceed the *breakdown voltage*. If that happens, the diode will maintain an almost constant voltage, that is, a zero incremental resistance. This range is not normally used with ordinary diodes, which have breakdown voltages in the neighborhood of 75 V, but is used to great advantage in the zener diode, which has a very predictable and usable breakdown voltage.

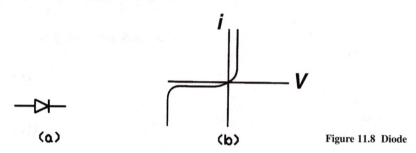

(a)   (b)   **Figure 11.8 Diode**

A simple application of a diode in a computing circuit is shown in Figure 11.9. Whenever V_{out} goes above the forward voltage drop of the diode (i.e., when the input voltage is negative), the diode will conduct, putting a near-zero resistance in the feedback path. This drops the gain of the amplifier to zero, so no further output voltage increases are possible, regardless of the input voltage. When the output voltage is negative, the diode stays in its nonconducting state, so the amplifier behaves normally. The presence of the diode in this instance thus yields a circuit that will act as an inverting gain with an output nominally limited to being negative only. Reversing the diode will give a similar circuit, with its output limited to positive values only.

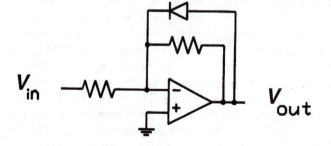

V_{in} V_{out} **Figure 11.9 Gain with a Diode Clamp**

Figure 11.10 extends the clamping circuit to clamping at an arbitrary voltage. The voltage divider will leave the diode reverse biased until the output voltage gets high enough to make the voltage at the wiper become positive. The diode then begins to conduct and the output voltage will not rise any further. This method can be applied to limit the output voltage on the high and low sides, if desired, by using a reversed diode and the opposite voltage supply for the second clamp. The circuit would then have arbitrarily set limits for high and low output.

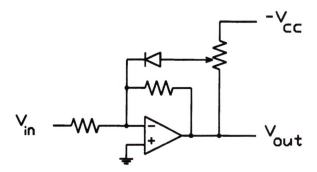

Figure 11.10 Gain with a Diode Clamp as Set Voltage

The use of diodes for clamping is extended to an absolute value circuit in Figure 11.11. It also functions as a full wave rectifier. The limit circuits above are half wave rectifiers in that when excited by a sine wave, the output will only show either the positive or the negative lobe, depending on the configuration of the diode. The absolute value circuit works by summing twice the output of a half wave inverting stage with the original signal. Summing twice the half-wave output (which is also sign inverted) with the original signal cancels one lobe of the sine wave and replaces it with a lobe of opposite sign, Figure 11.12. The top signal is the original sine wave. The middle one is twice the negative of the half-wave rectification of the original signal, and the third is the negative of the sum of the top two signals, which is the desired result.

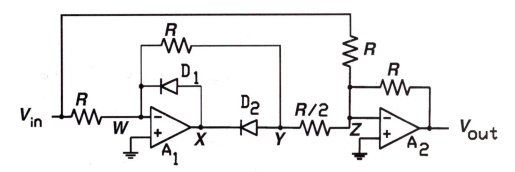

Figure 11.11 Absolute Value Circuit

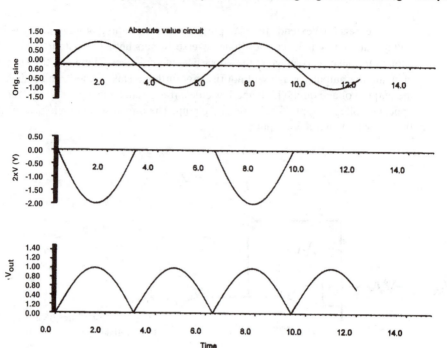

Figure 11.12 Signals in Absolute Value Circuit

The circuit is basically a combination of a limiter and a summer, with the addition of a series diode between them. Its operation can be traced by first noting that amplifier A1 is an inverting limiter. Whenever V_{in} becomes negative, its output will attempt to go positive, but diode D_1 will start conducting, thereby clamping the output. The voltage at point X will thus be negative or zero only, with unity gain. As long as X is negative, diode D_2 will conduct, so the output of A_1 will appear at point Y as one of the inputs to amplifier A_2. In this phase, the output of A_1, with a gain of 2 (because of the $R/2$ input resistor), and the original signal are summed (and inverted). The output voltage during this phase will be positive, with the same amplitude as the input, which is also positive [keep in mind that the diodes are conducting when V_{in} is positive (i.e., the output of A_1 is negative)].

During the second phase, the input voltage becomes negative. Both diodes enter their nonconducting modes. Diode D_2 serves to isolate amplifier A_1 from amplifier A_2. Diode D_1 serves to keep the summing junction of A_1 at its virtual ground state. With the two amplifiers isolated, A_2 acts as a unity-gain inverter for the input voltage. The input voltage at this time is negative, so the output is positive with the same amplitude.
Diode D_1 is necessary so that the voltage at point W, the summing junction of A_1, stays near ground. That guarantees that no current flows from point W to point Z during the isolation phase, which is essential for the proper operation of A_2 during this phase. With the addition of a low pass filter following this circuit an ac-to-dc converter can be realized.

11.2.5 Comparator

Conversion of signals to standard digital ranges is often important when analog and digital systems coexist. A comparator is most often required to produce an output that will be used for later digital processing. The simplest comparator circuit is to use an operational amplifier with no feedback at all, with one of the voltages to be compared connected to the + input and the other to the - input. It will swing between its output voltage limits, depending on which of the input voltages is larger. This has two potential problems: the voltage limits may not match the digital requirements, and the amplifier will always be in saturation, which can require some time for recovery.

An inverting configuration can be used to solve both of these problems by providing diode limits on the output voltage. Figure 11.13 shows a circuit that might be used to interface to TTL elements. Diode D_2 prevents the output from becoming negative. Diode D_1 and its voltage divider limit the output voltage to 5 V on the positive side. A disadvantage of this circuit is that one of the inputs must be inverted before the comparison. Scaling can be combined with comparison by using different input resistors. That works in this circuit because the circuit is actually sensitive to the sign of the *current* input to the summing junction.

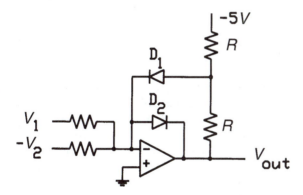

Figure 11.13 Comparator with TTL Output

Figure 11.14 shows a modification that makes the comparator behave like a Schmitt trigger. This is accomplished by the addition of some positive feedback via the noninverting (+) input. This adds a hysteresis to the switching so that small amounts of noise in the input do not cause the comparator to make many switches as the crossover point is reached. The positive feedback can also improve the switching speed somewhat.

When the output is at ground (input sum is positive; $V_1 > V_2$), the voltage at point A is the same as at point B, so no current flows. In this case the amplifier will switch when $V_1 - V_2$ crosses zero (so that $V_1 < V_2$), causing the output to go to +5 V. On the reverse switch, with V_1 going from less than V_2 to greater than V_2, the voltage at point A will be 5 V, so that the voltage at B will be positive. In this case, $V_1 - V_2$ will have to overcome the positive voltage at B before the amplifier can switch, providing the desired hysteresis. If the positive feedback has been added to speed the response, the hysteresis may be considered undesirable.

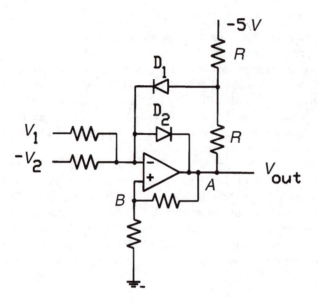

Figure 11.14 Comparator with Positive Feedback

11.3 FOLLOWER CONFIGURATIONS

Most of the applications given so far have used the inverting input (-) for the signal input, with the noninverting input (+) grounded. The noninverting input can also be used, with somewhat different but often desirable results. Figure 11.15 shows a noninverting amplifier that provides a gain. The circuit's ideal characteristics can be derived from the same rules as those used for the inverting circuits. Because of the infinite input impedance, the amplifier draws no current. The current flow from the output thus all goes through point P to ground. The voltage at P is

$$V_P = \frac{R_1}{R_1 + R_2} V_{out} \tag{11.16}$$

Because the amplifier ideally has infinite gain, the difference between must be negligibly small, so

$$V_{in} = V_P \tag{11.17}$$

and therefore,

$$V_{out} = \left(1 + \frac{R_2}{R_1}\right) V_{in} \tag{11.18}$$

This amplifier has some advantages and disadvantages with respect to its inverting counterpart. First, it does not invert. Also, it has the advantage of presenting the amplifier's very high input impedance (ideally infinite) to the source. An inverting amplifier has an

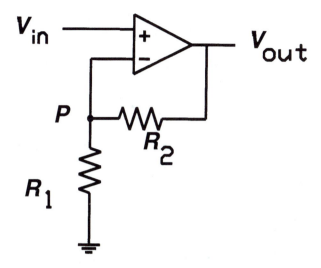

Figure 11.15 Noninverting Amplifier

input impedance determined by the input resistor, which is much lower than the input impedance of a typical operational amplifier. Its disadvantage is that the form of the gain relationship allows for much less flexibility in the generation of interesting dynamic functions because of the "1" that appears. Also, this amplifier is a true voltage amplifier, so it is not as easy to build a summing amplifier as it is with an inverting configuration, whose behavior depends on current summations.

An interesting amplifier configuration results from letting $R1$ go to infinity and $R2$ go to zero (Figure 11.16). With these values substituted in equation (11.18), the output voltage is

$$V_{out} = V_{in} \tag{11.19}$$

This configuration is known as a *voltage follower*. The output voltage is equal to the input, but this amplifier does far more than a wire. With an ideal operational amplifier, the input current is zero and the output current can be anything the load requires. Ideally, it thus acts as a perfect isolator. If the source is an instrument, for example, with very low current capability (high output impedance), using a voltage follower will allow further

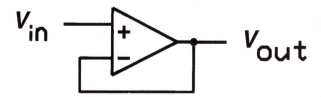

Figure 11.16 Unit Follower

processing of the signal without fear of drawing too much current from the instrument and thereby changing the voltage.

Noninverting configurations can also be used to build nonlinear circuits. Figure 11.17 shows a circuit that is a peak follower/holder. As long as the charge on the capacitor is not disturbed, the circuit output will always give the highest voltage that has been reached by the input so far. As long as the input voltage is higher than the capacitor's voltage, the diode will be forward biased, and the capacitor will charge up to the level of the input voltage. The second-stage amplifier is just a follower, so it will always track the capacitor's voltage. If the input voltage decreases, the diode will be reverse biased and will stop conducting, disconnecting the capacitor from the input. The capacitor will continue to hold its last voltage, and that value will appear at the output.

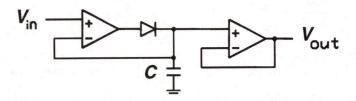

Figure 11.17 Peak Holder

A passive circuit, with just the diode and a capacitor, will also act as a peak detector, but this circuit has much better behavior. The two major problems with the passive peak holder are that the current to charge the capacitor must be drawn from the source, the capacitor voltage will only be within one diode drop of the actual peak, and the capacitor will leak, so the result will only be valid for a short time.

The operational amplifier circuit solves these problems. The first-stage voltage follower guarantees that very little current is drawn from the source. The feedback from the capacitor solves the second problem. If the capacitor voltage is below the input voltage, a voltage difference will be generated at the input to the first amplifier. This will generate a large voltage, which will overcome the diode's forward drop and charge the capacitor. As long as the peak voltage is held for long enough to charge the capacitor, the difference between the capacitor's voltage and the input voltage at a peak will be inversely proportional to the amplifier's open-loop gain, which is very large. The diode's forward drop therefore does not figure in the system's behavior.

This selection represents only a small fraction of the potential applications of operational amplifiers. They are one of the most versatile tools available to designers and is essential to the successful operation of most mechanical control systems.

11.4 DIGITAL-TO-ANALOG CONVERTERS

As shown in Chapter 9, the summation circuit for a D/A converter can be built by using an operational amplifier as an inverting summer (Figure 11.18). The circuit shown in

Figure 11.18a sums the reference voltage with unit gain, and other voltages with gains that are successively doubled. The reference voltage is negative, so the output will have a positive range. As shown, the circuit is a three-bit converter with nominal range 0 to 10 V. The actual maximum output voltage is 8.75 V for an input word of 111 (1.25 + 2.5 + 5); the 10 V output would be produced by an input of 1000. As shown, the input word is 100 for an output of 5 V.

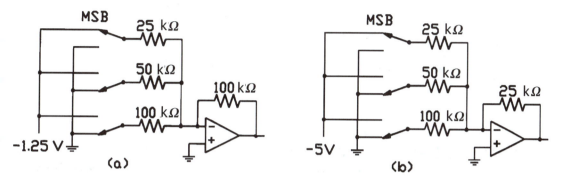

Figure 11.18 D/A Converter Using Summation

The circuit in Figure 11.18b is a small variant on the original circuit. It uses unity gain for the highest-order bit instead of the lowest-order bit and uses a –5-V reference. The main advantage to this circuit is that it is usually easier to achieve an accurate reference voltage at the higher voltage level. This summation circuit, although it is an excellent illustration of a D/A converter, has serious shortcomings. The most important of these is that when the precision is expanded, a very wide range of resistor values is needed. It may not even be possible to get the needed resistors from the set of standard resistor values. Even neglecting the availability problem, the large ratio of resistor values (4000:1 in a 12-bit converter) might get the operational amplifier out of its ideal range. Furthermore, each resistor would have different temperature-sensitivity characteristics, so the calibration of the converter would not be very stable.

Figure 11.19 shows an alternative circuit that circumvents these problems. This circuit uses a follower configuration, so does not sum the input voltages. Instead, each combination of the switch positions gives a different series-parallel configuration of the resistors, which can be treated as a resistor network and, ultimately, as a voltage divider to find the voltage at the V_+ input. Because the operational amplifier ideally draws no input current, its input voltage depends only on the resistor network. The amplifier itself acts as a follower to isolate the input voltage. This circuit, called a *ladder circuit* because of the arrangement of the resistors, uses only two unique resistor values, regardless of the precision of the converter. Like the summation converter, the nominal output range of the converter is 0 to 10 V, with the actual maximum output equal to 8.75 V. In the configuration shown, the input is 100, giving an output of 5 V. This output voltage can be derived by looking at the equivalent input circuit for an input word of 100 (Figure 11.20). If the equivalent resistance of the part of the network to the right of the V_+ point is computed, it is found to be 20 kΩ, so V_+ is halfway between the reference voltage and ground. Each switch configuration generates a different equivalent circuit. If bipolar output is desired, a negative voltage source can be substituted for the ground.

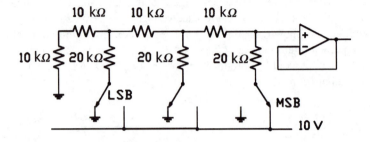

Figure 11.19 D/A Converter Using Ladder Network

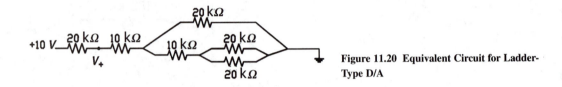

Figure 11.20 Equivalent Circuit for Ladder-Type D/A

11.5 FUNCTIONAL CHARACTERISTICS

The development of typical operational amplifier circuits was based on properties of the ideal amplifier:

- Infinite open-loop gain
- Infinite input impedance (draws no current)
- Zero output impedance (voltage independent of load)

The pleasure in using operational amplifiers is that for most applications they are close enough to these ideals so that the circuits work essentially as expected. When pushed, however, the real properties take over and determine the actual system performance.

In addition to the ideal properties listed above, the actual operating characteristics can be referenced to an expanded list of ideal characteristics:

- Infinite open-loop gain
- Infinite input impedance (draws no current)
- Zero output impedance (voltage independent of load)
- Zero offset voltage
- Infinite bandwidth
- Zero response time
- Zero variation of characteristics with common-mode input voltage
- Zero variation of characteristics with power supply voltage
- Zero variation of characteristics with temperature

11.5.1 741 Op-Amp

The 741 isn't an airplane—it is the part number for by far the most popular operational amplifier. 741s are very inexpensive and available from many manufacturers. In most applications, if an operational amplifier is used and its type is not mentioned, it is probably a 741. There are also a wide variety of other operational amplifiers available. They are designed from various technologies and for many special applications. Various grades are available for quality selection over most of the amplifier's properties and environmental tolerances.

11.5.2 Open-Loop Gain

The open-loop gains available range over:

- 10^4 (80 dB) for standard industrial-grade amplifiers
- 10^5 (100dB) for readily available amplifiers
- 10^6 (120dB) for expensive amplifiers

The dB values are expressions for the gain using a logarithmic unit defined by

$$\text{gain(dB)} = 20 \log_{10} (\text{gain}) \qquad\qquad (11.20)$$

The gain determines how close to ground the summing junction (virtual ground) stays. At lower gains, the summing junction voltage rises and introduces some degree of amplifier characteristics into the input–output functions.

11.5.3 Input Impedance

The input impedance varies with the technology used in constructing the operational amplifier. Typical values are:

Type	Input impedance
Transistor input	300kΩ to 3 MΩ
Darlington	1 to 10 MΩ
FET input	100 MΩ to 1000 GΩ

The input impedance controls the current that the source will be required to supply to operate the amplifier. In follower configurations, the source sees the input impedance directly. In computing (inverting) configurations, the input resistance determines the impedance seen by the source. The high limit on the input impedance, however, is controlled by the amplifier input impedance. The current flow through input resistance must be much larger than the current drawn by the amplifier so that the assumption of zero current flow will hold. If the input resistance is too high, the input current flow will fall to levels such that the assumption is no longer valid.

11.5.4 Output Impedance

Typical output impedance values are 150 to 200 Ω. The range of values can be from 10 Ω to 5kΩ. The output impedance puts restrictions on the types of loads that can be connected to the amplifier. For the output voltage of the amplifier not to be affected by the load, the input impedance of the load must be much higher than the output impedance of the amplifier. Looking at the input impedances typical for operational amplifiers (above), it can be seen that connecting the output of one amplifier to the input of another preserves this relationship.

11.5.5 Input Offset Voltage

When the inputs of an amplifier are both grounded, there will normally be a small output voltage. Typical values are:

Military grade	1 mV
Industrial grade	3 mV
Consumer grade	10 mV
FET types	10 X's as large

This small output voltage usually has little effect on systems of gains and other static components but can wreak havoc in systems with integrators, which will patiently accumulate charge even when the inputs are zero. Internal or external nulling circuits can reduce this drift, as can chopper stabilization.

11.5.6 Bandwidth

As the frequency of the input signals rises, the open-loop gain of the amplifier falls. The bandwidth property is often expressed as the gain–bandwidth product (GBP), with the bandwidth defined as the frequency at which the gain drops by 3 dB from its open-loop value (a drop to about 70% of its original value). The gain–bandwidth ratio is typically 1 to 10 MHz. With a gain of 10^5, for example, a GBP of 5 MHz would yield the very modest bandwidth of 50 Hz. This should not be taken as too alarming, however, since this represents the frequency at which the gain is 70% of its dc value. As long as the gain stays high enough for the infinite gain assumption to hold, the circuit function will not be affected significantly by the actual gain value.

11.6 PROBLEMS AND DISCUSSION TOPICS

1. Assume that the circuit shown in Figure 11.21 operates from a supply voltage of V_{cc}. With the resistor values as shown, determine the saturation level at V_{out} and draw the relation between V_{out} and V_{in}, assuming an ideal diode and an ideal op-amp.

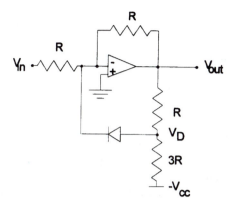

Figure 11.21 Diode Limited Amplifier

2. The basic noninverting amplifier circuit is shown in Figure 11.22. Assume that the input impedance of the amplifier is infinite but the open-loop gain is a large finite value, A. Derive the relation between V_{out} and V_{in} in terms of the gain and resistance values. Verify that as A tends
toward infinity, the result reduces to the result of equation (11.18).

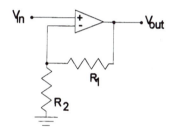

Figure 11.22   Noninverting Amplifier

3. Set up a dc motor with an analog tachometer.

   (a) Build an op-amp circuit (in computing mode) to implement a P (proportional) speed control. Find the tuning that achieves best performance. Set up a computer control as well
   and compare the maximum attainable gains (after accounting for scaling and units) and
   the associated performances.

   (b) Add a frictional load so that the P control will show significant steady-state error (if a
   voltage amplifier is used, there could be significant steady-state error without the frictional load). Redesign the controller so that it becomes a PI (proportional plus integral)
   controller.

4. Set up a dc motor with an analog tachomoter and an incremental encoder. Use a computer and
decoder chip to record position. Use an op-amp integrator to attempt to estimate position from
the velocity (tachometer) signal. Read the estimated signal into the computer, scale the results,
and compare to the encoder measurement. Discuss the results and whether or not this is a suitable way to estimate position.

5. Create an analog signal containing a mixture of a 10-Hz sine wave and a 1000-Hz sine wave.
Attempt to capture that signal using a computer data acquisition system sampling at 100 sam-
ples/sec. Show the effects of aliasing. Using op-amps in computing mode, construct an
antialiasing filter and tune the cutoff frequency for "best" results (define what is meant by
"best").

6. Using the gain–bandwidth specification for an op-amp, set up a computing circuit as a unit

gain, and show that the performance becomes measurably different from ideal as the frequency of the input signal increases. Compare the observed and predicted frequency at which such a loss of gain takes place.

7. A low-powered instrument type of interface can be simulated by connecting a voltage source through a large resistor. Set up such a source, then use both computing (inverting) and follower (noninverting) op-amp unity-gain amplifiers to measure the deviation of the performance from the ideal as the source resistance is increased. Compare the performance of the two op-amp configurations.

8. When building control systems that include an integration (I) term, it is often necessary to limit the integral output to some value less than the controller output limitation. Design an op-amp circuit for a PI control which has independently adjustable limits for the integral term and the controller output. Test it on an appropriate physical system (e.g., a motor with voltage amplifier or a frictional load, temperature control of a thermal system, etc.). Note that some systems have asymmetric limits, such as a thermal system with heating only.

9. Build a motor-driven system with analog velocity and position instruments. Design an op-amp circuit to estimate the velocity from the position instrument and compare its estimate to the velocity measurement from the velocity instrument. Build a PID position controller, first using the velocity instrument for the D term and then using the estimated velocity. Compare the performance.

12

Power Amplifiers

The ultimate goal of any mechanical system control is to deliver power. Because of the broad use of electrically based information throughout control systems, it is convenient if the final power element is an electric–mechanical–thermal/etc. power conversion unit. The single most common such unit is the electric motor in all its variants. This chapter is concerned with how an electrical command signal is used to modulate the power delivered by these power conversion systems.

12.1 INTRODUCTION

12.1.1 Overview

In this chapter we highlight the essential issues associated with power amplifiers that are often encountered in the design of mechanical systems, specifically, amplifiers providing dc-coupled power ranging from a few watts to roughly a kilowatt. Typical applications include fractional horsepower dc motors, piezoelectric transducers, and solid-state heaters, to name just a few. The information provided here is aimed at the control or mechanical engineer, who must design a system that includes a power amplifier; this is intended to assist in the design or selection of an appropriate device. The detailed design and construction hints included here are oriented toward the construction of a single or perhaps a small quantity of units. In these circumstances, unit cost and hand tuning are regarded as a small price to pay for robust performance. For the engineer who may be faced with the task of either specifying the purchase or the design of an amplifier for a large production run, this information should still be of value in specifying systems or working with power system designers.

A small amount of background material is essential. Knowledge of the following areas from basic electronics is assumed:

- *R L C* circuits
- How a diode works
- How an "ideal" op-amp behaves
- How logic gates function

Furthermore, familiarity with the following notions from classical controls is assumed in some sections:

- Block diagrams
- Root-locus design techniques
- Frequency response and Bode plots

The chapter is organized in the following manner: A short introduction to the basic notions behind power amplifiers is provided, followed by a short overview of power transistors. Then more detailed information on linear, pulse-width-modulated (PWM), and integrated-circuit power amps are given. Finally, hints on integrating power amplifiers into control systems and a few tips on construction and safeguards are offered.

12.1.2 Voltage Output versus Current Output

The role of the power amplifier is to boost the power level of a signal. An ordinary op-amp can deliver only a limited amount of voltage and current to a load. Typical limits are on the order of 20 V and 25 mA. Since this is insufficient for most electromechanical actuators, some means of boosting the power is required. For some of the types of actuators mentioned above, currents on the order of amperes or tens of amperes may be required, and others may require voltages on the order of 100 V or more. There is no sharp demarcation, but the term *power amplifier* usually refers to devices that have output voltages on the order of 20 V or more and/or output currents on the order of 100 mA or more.

The input (command) signal to a power amplifier is usually a voltage. Use of current as the input signal is certainly valid but is not encountered as frequently. The remainder of the notes will consider cases where the input signal is a voltage. The output variable of interest can be either voltage or current. If the input is a voltage and the output is specified in terms of a voltage, the amplifier is simply termed a *voltage amplifier*. Implicit in this definition is the fact that the amplifier will produce whatever output current is necessary to maintain the desired voltage gain. If the input is a voltage and the output is specified in terms of a current, the amplifier is termed a *transconductance* or *current amplifier*. Here, it is implicit that the amplifier will produce whatever output voltage necessary to maintain the desired current (transconductance).

Consider the voltage amplifier in more detail. There will be a prescribed gain (expressed in V/V) which can be regarded as constant at low frequencies and will eventually drop off at high frequencies due to small parasitic capacitances and resistances present within the electronic devices. The amplifier maintains the desired input–output voltage ratio (the voltage gain) and lets the load draw whatever current necessary. Realistically, this current cannot become infinite and will begin to clip at the maximum current that the amplifier can deliver. This limit is up to the discretion of the designer and is a design choice based on the quantity and quality of power transistors in the output stage of the amplifier. The specifics of such design issues are covered later. To emphasize the basic idea, consider the example shown in Figure 12.1. The values here are representative of a highly simplified model of a fractional-horsepower dc motor. Assume that the load is adequately modeled as a simple resistor, R_1. The amplifier is represented by the familiar triangular element used for regular op-amps. Assume that the amplifier has a voltage gain of 5 V per volt. Let the maximum voltage output of the amplifier be within 1 V of the supply rails, or ±34 V. Assume that the output current is limited to ±10 A.

Applying a series of dc voltage signals, V_i results in a corresponding series of output voltages and currents, V_o and I_o. Consider the system with R_1 equal to 4 Ω and then repeat the same series of inputs with R_1 equal to 2 Ω. The resulting signal values are summarized in Table 12.1. Now consider the transconductance amplifier. There will be a prescribed gain (expressed in A/V) which will again be constant at low frequencies and will eventually drop off at high frequencies. For this case the amplifier maintains the input–output transconductance specified by the gain and allows the voltage across the load to be as large as necessary, up to the maximum voltage the amplifier can deliver. As an example of this, use the same values for load resistance and output voltage limitations used in the example above. In this case let the transconductance of the amplifier be 2.5 A/V. This is shown in Figure 12.2 and the operating conditions are given in Table 12.2.

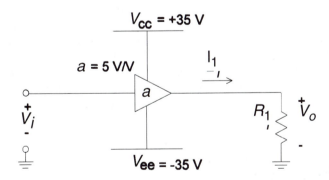

Figure 12.1 Simple Voltage Amplifier

TABLE 12.1 VOLTAGE AMPLIFIER OPERATING CONDITIONS

R_1 (Ω)	V_i (v)	V_o (v)	I_0 (A)	Comments
4	1.0	5.0	1.25	Normal operation
4	5.0	25.0	6.25	Normal operation
4	10.0	34.0	8.5	Voltage saturation
2	1.0	5.0	2.5	Normal operation
2	5.0	20.0	10.0	Current saturation

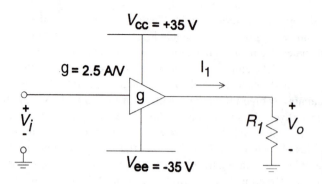

Figure 12.2 Simple Transconductance (Current) Amplifier

TABLE 12.2 TRANSCONDUCTANCE AMPLIFIER OPERATING CONDITIONS

R_1 (Ω)	V_i (V)	I_o (A)	V_o (V)	Comments
4	1.0	2.5	10.0	Normal operation
4	3.0	7.5	30.0	Normal operation
4	4.0	8.5	34.0	Voltage saturation
2	1.0	2.5	5.0	Normal operation
2	5.0	10.0	20.0	Current saturation

       The key idea is that the output of the amplifier may be specified as either a voltage (in which case the load will determine the output current) or a current (in which case the load determines the output voltage). The voltage and current are always limited and the amplifier will saturate when either voltage or current reaches a limiting value, regardless of whether it is a voltage or transconductance amplifier. Furthermore, in the overall design of a control system, some of the undesirable but unavoidable properties of the amplifier may need to be taken into account. While the ideal amplifier will have a frequency response that is constant, real amplifiers have limited bandwidths. Real amplifiers do go into saturation as the operating limits are reached. In addition, amplifiers may exhibit more subtle vices, such as a small deadband near the zero output level, small offsets, mild nonlinearities, or injection of noise into the system. The designer of the mechanical system must evaluate how this will affect the system and take steps to eliminate the problem without introducing undue complexity.

12.1.3 Linear versus PWM Amplifiers

The two main types of power amplifiers are (1) linear and (2) pulse-width modulated (PWM). The choice between the two is dependent on the particular application, the main factors being cost, efficiency, and performance. In a linear (or proportional) amplifier the output signal can take on, at any given instant in time, any value within the saturation limits. It is truly an analog device. The amplifier is linear in the sense that the output voltage or current is a linear function (i.e., related by a constant gain) of the input voltage. Modeling the power amplifier as a linear element introduces only minimal errors. The frequency response of the amplifier is normally designed so that the bandwidth of the amplifier is much greater than the bandwidth of the actuator.

Aside from keeping system design simple, some applications demand the precision and "purity" of a truly linear amplifier. An obvious example is in an audio system. Any spurious harmonic overtones or subharmonics arising from nonlinearities would be unacceptable. In situations where very high bandwidth or very low noise is required, a linear amplifier is necessary. The disadvantage of a linear amplifier is power inefficiency. Linear amplifiers dissipate power within the output stages of the amplifier itself. This means that they consume more power, must be larger and more robust from a thermal standpoint, and can generate a lot of heat.

In a PWM amplifier the output signal can take on, at any given instant in time, only one of two values. The output signal is a periodic waveform with a constant frequency and a variable duty cycle. The duty cycle is modulated by the input voltage. The switching frequency is typically on the order of 20 kHz or more. The switching frequency is normally set well beyond the bandwidth of the actuator and the actuator acts as a low-pass filter. The output of the actuator will effectively be the average value of the PWM signal supplied by the amplifier. In this way, the output signal of the actuator can be adjusted to take on any range of values below the saturation limits of the amplifier and thus approximates a truly analog system. While the actuator acts as a low-pass filter, the harmonic content present in the amplifier's waveform does get passed through to the output. The amplitude is normally small, but in some applications this may be unacceptable. One of the motivations for putting the switching frequency in the 20 kHz range is so that any audible noise generated by the switching is beyond the range of human hearing.

The main benefit of this approach is that it minimizes the power dissipation within the amplifier itself. The key drawback is in the analysis of the system. While a simple averaging method will yield a linearized model of the switching amplifier, the level of accuracy may be insufficient for high-performance systems. The PWM amplifier is a fundamentally nonlinear device and obtaining an analytic model that is reasonably accurate is difficult at best. It should be noted that either voltage or transconductance amplifiers can be built as either linear or PWM amplifiers. To emphasize the distinctions between a linear and a PWM power amp, consider a voltage amplifier with a gain of 4 V per volt. Assume that both the linear and the PWM amplifier operate from supply rails of ±25 V. Assuming that no saturation occurs, the waveforms shown in Figures 12.3 through 12.7 illustrate the input and output waveforms.

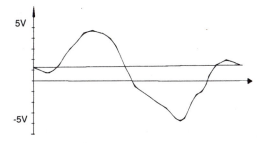

Figure 12.3 Input Signal versus Time

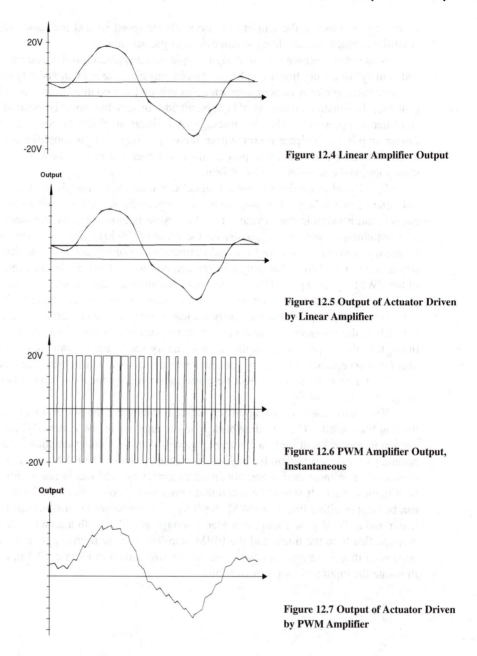

Figure 12.4 Linear Amplifier Output

Figure 12.5 Output of Actuator Driven by Linear Amplifier

Figure 12.6 PWM Amplifier Output, Instantaneous

Figure 12.7 Output of Actuator Driven by PWM Amplifier

12.2 POWER TRANSISTORS

12.2.1 Background

Power transistors are the building blocks used in the construction of power amplifiers. They are distinguished from ordinary transistors by their higher operating limits in terms of voltage and current. The basic principles of operation are essentially the same as for ordinary transistors. The objective here is to review the operational behavior of transistors very quickly and then move quickly on to the distinctions between power transistors and regular transistors. Both bipolar junction transistors (BJTs) and metal-oxide semiconductor field-effect transistors (MOSFETs) will be considered. The review will be largely qualitative; a more quantitative discussion can be found in a large number of texts dealing with integrated-circuit design.

Power transistors are a small subset of the more broad field of power semiconductors. Many special-purpose devices are currently available on the market and the number is increasing rapidly. While intimidating at first glance, most of these devices are outgrowths of more familiar devices: diodes and transistors. To introduce the transistor, regardless of whether it is a BJT or a MOSFET, think of it as mechanism for controlling the flow of current. The transistor is a three-terminal device; one where the main current flows in, one where the main current flows out, and a third for the application of a controlling signal. Depending upon the device, a very small amount of current may flow into or out of the control terminal.

12.2.2 Bipolar Transistors

Bipolar junction transistors (or simply bipolar transistors) come in two varieties, the NPN and the PNP. Figure 12.8 shows the symbol for each, along with the name of each terminal and the direction of current flow for normal operation. Consider first the NPN transistor. The main current flows in at the collector and flows out at the emitter. The base is where a controlling signal is applied. Note that the path from the base to the emitter in the schematic has an arrow, which is reminiscent of a diode. This is not a coincidence; when the transistor is in normal operation there will be voltage drop of about 0.7 V (one diode drop) from the base to emitter. More precisely, the collector current, I_c is an exponential function of the base–emitter voltage, V_{be}. For small values of V_{be}, below about $0.6V_c$, I is negligible. As V_{be} is increased, I_c rises sharply.

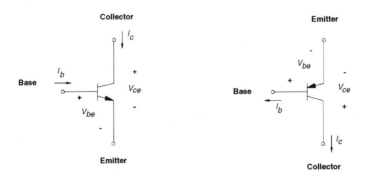

Figure 12.8 Bipolar Transistors

For the output current levels of interest (say I_c equal to tens of milliamperes to a few amperes) there will be slight variations in the V_{be}, but 0.7 V is a good approximation. Some current will be drawn into the base. The relation between the base current and the collector current is very well approximated by the relation $I_c = \beta I_b$. Consequently, specifying a V_{be} or an I_b for a transistor acts as the controlling signal that will determine the main current flow, I_c. The current flowing out, I_b, is the sum of I_c and I_b. Since I_b is usually much smaller than I_c, taking I_e as being equal to I_c is a reasonable simplification.

Textbooks usually use the constant ß, whereas manufacturers' databooks will often use h, hf, or h_{fe}. Regardless of the nomenclature, this parameter is typically between 20 and 200. Unfortunately, the tolerance from one transistor to the next is enormous, on the order of 50% or more. Manufacturers' databooks usually give minimum and typical values. At any rate, a good design will not depend on the absolute value of ß. The voltage across the collector–emitter junction, V_{ce}, does not exhibit a simple constant relationship as in the case of V_{be}. Clearly, for current to flow from collector to emitter, V_{ce} must be positive. In fact, for the transistor to work properly, V_{ce} must be above a minimum level, typically 0.2 V. The maximum possible value of V_{ce} is limited by the construction of the particular transistor. This is one distinguishing feature of a power transistor, as it is usually at least 35 V.

The PNP transistor functions essentially like the NPN transistor, with the main distinction being the polarity. Again, V_{be} is essentially constant at 0.7 V. Also, the main current flow is from emitter to collector and current flows out of the base. Again, V_{ce} will be between roughly 0.2 V and the limit for the device, at least 35 V. Minor differences associated with a PNP is that the ß is typically half that of an NPN. The tolerance is still huge.

One popular method of raising the effective ß in BJTs, particularly in power transistors, is to use what is called a *Darlington configuration*. It is a cascade of two transistors as shown in Figure 12.9. The example shown is for NPN transistors; the Darlington configuration is equally popular in PNPs. From the user's perspective, the Darlington is still a three-terminal device that exhibits the same operating characteristics as an ordinary transistor except that the effective ß is roughly the square of the ß of an individual transistor and that the effective V_{be} is two diode drops. Darlington configurations exist for PNP transistors. To boost the speed of the device, a complementary version that uses an NPN transistor is often used. When using this version, note that there is only one V_{be} between the base and emitter of the composite device. Both the traditional and complementary Darlington devices are shown in Figure 12.10.

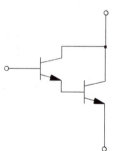

Figure 12.9 NPN Darlington

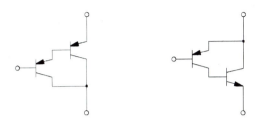

Figure 12.10 PNP Darlington

Having reviewed the basic operaton of BJTs, some distinctions should be made about the power semiconductors. Unlike the transistors that make up op-amps (which are subjected to relatively small changes in operating point), power transistors are typically used over a wide range of voltages and currents. Consequently, parameters (such as ß) will fluctuate. A good design will not be overly sensitive to these widely varying parameters. For example, as in signal-level op-amps, feedback can often be used to force the performance of a circuit to be dominated by resistor values rather than transistor parameters. Moreover, it is important to evaluate the kind of operating conditions that the transistor is likely to encounter.

Manufacturers' databooks will specify the maximum current (largest I_c) for the device and the maximum voltage (largest V_{ce}) that a device can handle. They also give a *safe operating area* curve that plots the allowable I_c versus V_{ce} at various temperatures and for intermittent operation. In addition to all the technical data, there are often detailed hints on mounting the devices to heat sinks. Many power transistors are designed with the expectation that they will drive inductive loads. More will be said about the ramifications of this later; for now, note that this can cause the load to apply very high voltages (often referred to as inductive kickback) to the transistors. To prevent destruction of the transistor, protection diodes are often built into the transistor, as shown in Figure 12.11. These are normally shown in the databook if they exist. If the amplifier will be driving an inductive load and they are not built into the output transistors, they should be added as a separate component. This property is one of the few disadvantages of solid-state power devices relative to the vacuum-tube devices that they replaced; vacuum-tube devices are much more resistant to damage from inductive loading.

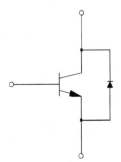

Figure 12.11 Protection Diodes

Finally, one big consideration with bipolar transistors is the effect of temperature. All BJTs exhibit a V_{be} that drops with increasing temperature. In other words, if V_{be} is held constant by external circuitry and the temperature is increased, I_c will increase. As

I_c increases, the power dissipated in the device increases and the temperature tends to rise. This condition, known as *thermal runaway*, leads to the rapid and often dramatic destruction of the power transistors in an amplifier. Fortunately, several design steps can be taken to reduce the likelihood of an amplifier going into thermal runaway.

12.2.3 MOSFET Transistors

MOSFET transistors also come in two basic types, the N-channel and the P-channel as shown in Figure 12.12. MOSFETS are one type of the more broad category of field-effect transistors (FETs). Further, within the MOSFET family there are *enhancement-mode* and *depletion-mode* devices. In addition, different manufacturers use different acronyms for devices that may be identical or have only very minor differences. Overall, this makes it difficult to associate the many names and many devices. The discussion here is limited to the fundamental devices, the enhancement-mode MOSFETs.

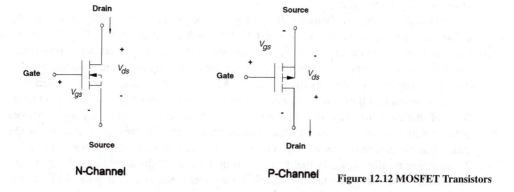

N-Channel   P-Channel   **Figure 12.12 MOSFET Transistors**

Like the BJT, the MOSFET acts as a device for controlling the flow of current. Consider first the N-channel device. The main current flows in at the drain and out at the source. The gate is the controlling terminal. The key distinction between a BJT and a MOSFET is that no dc current flows into the gate. It is like a BJT, but with ß raised to infinity. The gate voltage is the controlling signal. The fourth terminal of the MOSFET is the *body*. This is sometimes also referred to as the *bulk* or *backgate*. This terminal is normally tied to either the source or the lowest voltage level present in the circuit (highest voltage for a P-channel device). The effect of this terminal is secondary and is seldom a design consideration. In power transistors, this terminal is often connected to the source internally and the designer is then presented with a three-terminal device.

The controlling signal for the MOSFET is dependent upon the gate-to-source voltage, V_{gs}. Whereas the BJT exhibited an exponential relation between I_c and V_{be}, the MOSFET's I_d is dependent on the quantity $(V_{gs} - V_t)^2$. The parameter V_t is known as the *threshold* voltage and can be taken as constant for a particular device. The value of V_t is on the order of 1 V and is given in the databooks. The key thing to note is that the relation between current and voltage is now a parabolic relation. Unlike the BJT, the normal operation of a MOSFET cannot be characterized by a simple rule for the value of V_{gs} for a large range of currents. With these distinctions in place, it is sufficient to note that increasing V_{gs} increases the current, although not as rapidly as in a BJT.

   Whereas bipolar power transistors are often simply larger versions of ordinary bipolar transistors, power MOSFETS are somewhat different. Although this does not change the nature of the fundamental operating characteristics, it greatly extends the current and voltage ratings of the devices. Unfortunately, the various manufacturers of power MOSFETS each has given a name to its own set of minor design variations. Some of the more popular names for power MOSFETS include DFETS, TFETS, and HEXFET. Unlike bipolar transistors, MOSFETS do not appear in Darlington configurations. This emphasizes one of the advantages of a MOSFET: namely, the ease of driving the gate. The current required to drive the gate is transient: just enough to build up sufficient charge on the gate to develop the required voltage, which is normally very small. In addition, when a MOSFET is used as a switch, the effective resistance between the drain and source is very small, typically less than 1 Ω. These two advantages make MOSFETS attractive for applications where the transistors will be used as a switch (i.e., PWM amplifiers). A more subtle distinction between BJTs and MOSFETs is related to the thermal behavior. In general, for devices with equivalent specs, the MOSFET will be less prone to going into thermal runaway.

12.3 LINEAR AMPLIFIERS

12.3.1 Class B Output Stage

This discussion will be limited to the design of dc-coupled amplifiers capable of both sourcing and sinking current. While a great deal of information about amplifier design exists, much of it is directed at communication and audio systems in which a dc response is not required. In addition, the discussion here assumes that the bandwidth of the amplifier be relatively low, with a -3-dB bandwidth on the order of 50 kHz or less, realizing that for mechanical systems, bandwidths are rarely required to be greater than 1 kHz.

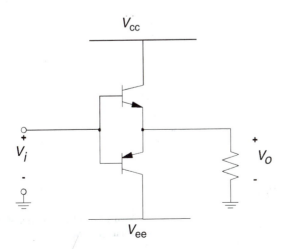

Figure 12.13 Class B Amplifier

The fundamental method used in simple output stages is the *push–pull* configuration shown in Figure 12.13. Electronics literature often classifies the various fundamental configurations by assigning a letter; the amplifier shown here corresponds to a *class B* output stage (class A refers to amplifiers for which the power transistor is always conducting; class B refers to configurations in which each transistor conducts over half the cycle). Most dc-coupled amplifiers are variants of class B stages, with some degree of modification in order to improve performance.

Consider an input voltage V_i equal to zero. In this case, neither transistor is on, so no current flows and the voltage across the load is zero, so V_1 is zero. Now let V_i increase and consider what happens to the NPN device. For Vi less than one V_{be} drop, negligible current is conducted and V_1 remains at zero. As V_i increases beyond one V_{be}, the NPN device begins conducting current and the PNP device remains off. The output voltage, V_1, follows V_i but is roughly one V_{be} drop less. This continues until the device is driven into saturation near the positive supply rail voltage, when V_{ce} reaches the saturation voltage of the transistor. Recall that this voltage is generally small, and using the supply voltage as the saturation voltage of the amplifier is a reasonable approximation. Note that the path of the current is from the positive supply rail and through the NPN device to the load.

Now consider the case when the input voltage swings in the negative direction. Lowering V_i to - V_{be} will turn on the PNP device and the NPN device will remain off. Further reductions will result in V_1 following V_i in the negative direction but with an increase of roughly one V_{be}. The path for the current is from ground through the load and then through the PNP device to the negative supply rail. The NPN device is off and has no effect. The input–output characteristics of the class B stage are shown in Figure 12.14. The obvious shortcoming is the *deadband* near the zero voltage level. This is also termed *crossover distortion*. In all but the most crude situations, this is unacceptable. Fortunately, there are a number of things that can be done.

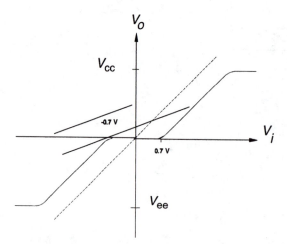

Figure 12.14 Class B Input–Output Characteristics

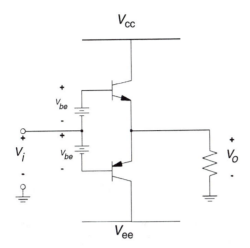

Figure 12.15 Class A-B with Ideal Compensation

12.3.2 Reducing Crossover Distortion

As pointed out in Figure 12.14, a significant amount of deviation from the ideal linear behavior exists in a class B output stage. The usual method for overcoming this is to add a bias voltage to compensate for the V_{be} drops of the output transistors. An idealization of this is shown in Figure 12.15. Trying to match the V_{be} values with perfect voltage sources will not work in practice, but fortunately, there are a number of things that can be done. Two of the most simple methods are shown in Figure 12.16. In the first scheme, a pair of diodes approximately match the V_{be} drops of the output transistors. Although this method is very simple, it has shortcomings. Precise matching of the diode drops to the transistor drops is generally not possible. If some crossover distortion remains, this is generally not a severe problem. If, on the other hand, the diode drops are too large, excessive current will flow in the amplifier in the quiescent condition, which can lead to problems with thermal stability. This can be mitigated somewhat by mounting the diode in thermal contact (adjacent to them, on the same heat sink) with the output transistors.

A better method is to use the configuration shown on the right. The combination of Q_3 and R_1 and R_2 forms what is called a V_{be} multiplier. The bias voltage is approximately $Vbe\ (1 + R_2/R_1)$. In a practical implementation, a single potentiometer is used for R_1 and R_2. This allows the quiescent current to be adjusted by changing the bias voltage. In addition, now only Q_3 needs to be mounted in thermal contact with the output transistors, which is generally much more convenient than mounting diodes.

12.3.3 Using Feedback within the Amplifier

The methods above are designed to eliminate the major sources of nonideal behavior. Despite the best design efforts, some nonlinearities will remain. Moreover, the frequency response, input impedance, and output impedance will be determined by the individual

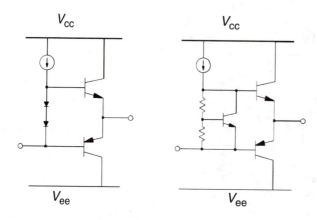

Figure 12.16 Practical Compensation Techniques

devices To stabilize the amplifier's characteristics in light of these uncertainties, feedback is almost invariably used. For most applications, an op-amp provides an ideal gain block and has sufficient gain–bandwidth product for most amplifiers destined for use in electro-mechanical systems. Fortunately, the simple rules that apply to designing ordinary op-amp circuits can be used as a first approximation when designing power amplifiers driven by op-amps. This is particularly true for conservatively designed op-amps, which have been compensated internally. As an example, consider a generic op-amp (such as a 741) used as shown in Figure 12.17.

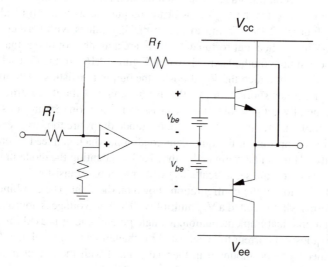

Figure 12.17 Voltage Amplifier with Internal Feedback

Since the voltage gain of the output stage is 1 (consider Figure 12.15 without the crossover dead zone), the analysis of this circuit reduces to the simple inverting amplifier, $V_o = -(R_f/R_i) V_i$. In some instances the output transistors will introduce enough parasitic capacitance and resistance to the circuit to reduce the phase margin of the amplifier dangerously when a lot of feedback is applied. In this case an op-amp that can be compensated externally would be used.

12.3.4 Thermal Stability

Thermal stability is an important design issue. The V_{be} of a bipolar transistor decreases with temperature. This can lead to *thermal runaway*, in which the V_{be} drop results in increased current. The increased current leads to further temperature increases and a further V_{be} drop. This typically results in destruction of one (occasionally both) output transistors. One way to guard against thermal self-destruction is to mount Q_3 in close thermal proximity to the output stages. Any changes that appear in the V_{be} of the output transistors will also appear in the bias voltages and the effects will largely cancel. As always, using a large heat sink and mounting devices carefully will assist the dissipation of heat.

Additional methods that help the thermal stability of the amplifier are to place power resistors in series with the emitters of the output transistors. This effectively raises the bias voltage required and results in additional power dissipation within the amplifier, but tends to make the amplifier less sensitive to changes in the V_{be} as the temperature changes. In addition, these resistors prevent a short circuit between the power rails in the event that both transistors are turned on at the same time. Finally, by adding diodes as shown in Figure 12.18, the output current of the amplifier can be clamped at a level given by $I_{max} = V_{be}/R$, where the forward voltage of the diode is assume to be equal to V_{be}. For the case of positive output voltage, when the current becomes sufficiently high,

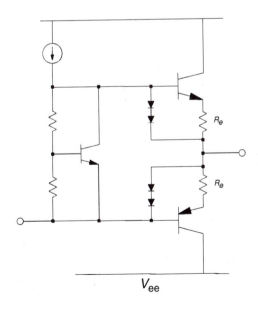

Figure 12.18 Techniques for Enhanced Thermal Stability and Current Limiting

the voltage drop across R causes D_1 and D_2 to turn on and clamp V_{be} at one diode drop. The same thing happens when D_3 and D_4 turn on when the output voltage is negative.

12.3.5 Variations on Class A-B Output

A method commonly used to boost the current drive capability of the class A-B amplifier is to add *booster* transistors to the output as shown in Figure 12.19. For low levels of positive output current, the voltage drop across R_1 is relatively small and the output current is provided by Q_1. As the output increases, the voltage across R_1 forces Q_3 to become active and deliver additional current. As the current through Q_3 becomes large, the voltage across R_2 becomes large. The purpose of D_2 is to match the V_{be} drop of Q_3. The diode D_1 clamps the output current delivered by Q_3 at roughly $0.7/R_2$. Resistors R_b and R_a form a local feedback loop to stabilize the voltage gain of the four output transistors at $(R_a + R_b)/R_a$. In the limiting case R_b can be set to zero and R_a can be infinite, yielding unity gain.

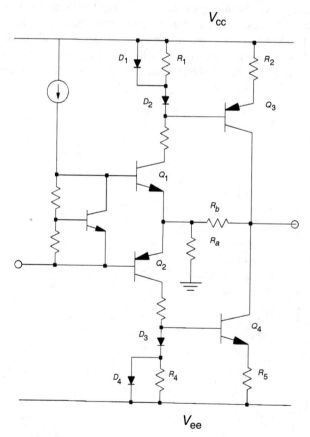

Figure 12.19 Techniques for Boosting Power and Current Limits

12.4 PWM AMPLIFIERS

12.4.1 Basic Operation

The key disadvantage of the linear power amplifier is the amount of power dissipated in the amplifier itself. In many instances, the desire for a more efficient amplifier outweighs the importance of linearity. In these situations, the pulse-width-modulated amplifier makes sense. In a PWM amplifier, the output transistors are either all the way on or all the way off. For a device that is all the way on, most of the voltage drop will occur across the load and very little will occur across the transistor, meaning that little power is wasted in the transistor. Similarly, little power is dissipated when the transistors are off.

The fundamental operation of any PWM amplifier can be explained by breaking the amplifier up into the four functional blocks shown in Figure 12.20. The timing generator is some sort of an oscillator running at the switching frequency. The comparator evaluates the input level of the amplifier as well as the current time in relation to the switching period and determines which polarity the output should take on. The signal conditioning makes small adjustments to the waveform timing and power levels to assure proper switching of the output transistors. The output transistors themselves form the final function block. Although the components used to make up each block will vary from amplifier to amplifier, each functional block will be present in some form.

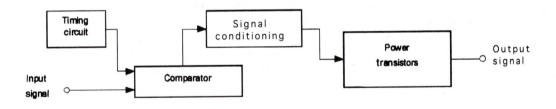

Figure 12.20 PWM Amplifier Block Diagram

12.4.2 PWM Input from an Analog Source

In some instances the command signal fed to the amplifier is analog. Several methods exist for the generation of a triangular waveform. Most are based on the idea of cascading a square wave into an integrator. Various databooks give examples. One technique that appears often uses a pair of op-amps as shown in Figure 12.21. Various extensions to this simple circuit exist in order to stabilize the amplitude and frequency of the output over supply voltage and temperature.

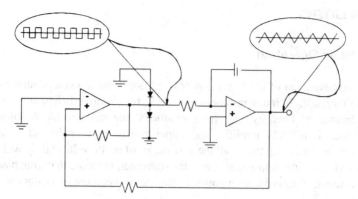

Figure 12.21 Sample Oscillator Circuit

The output of the comparator will generally require some signal conditioning. Output transistors do not necessarily turn on and off at the same rates. If an H-bridge output is used as shown in Figure 12.22, transistor pairs Q_1, Q_4 and Q_2, Q_3 are switched at opposite times. If Q_1, Q_4 turn off more slowly than Q_2, Q_3 turn on, the power supply rails are shorted briefly. The currents that flow in the transistors may become excessively large, leading to destruction of transistors. While emitter resistors could be added to limit the current, that defeats the purpose of going with the more efficient PWM amplifier. A better method is to insert a slight delay in the base drives of the transistors when they switch from low to high.

In addition to the fact that the output transistors may not turn on and off at the same rates, the switching may be too slow in general. This is often a result of the fact that the upstream device can not deliver sufficient current, particularly the large transients that are required at turn-on and turn-off. This is true even of MOSFETs. Smaller MOSFETS can often be driven directly from logic gates, but larger ones will have a relatively high capacitance at the gate that needs to be charged up until the desired V_{gs} becomes sufficiently large. The rate at which the gate voltage rises is dependent on the current flowing into the gate.

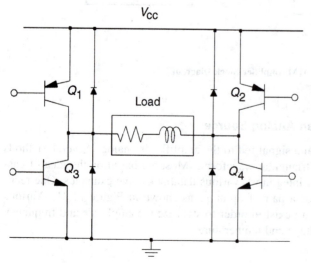

Figure 12.22 Complementary Bipolar H-Bridge

12.4.3 PWM Input from a Digital Source

In many control applications, the control algorithm is running on a microprocessor. For situations like this, feeding the output of the controller into the amplifier in digital format will improve the resolution and reduce the noise by eliminating the D/A conversion process and associated circuitry. In this case the timing generator is a crystal, which necessarily runs at a much higher frequency than the switching frequency. The reference level is maintained in digital form and at the beginning of each switching period and a counter is reset to zero. With each clock pulse the counter is incremented and compared to the reference value. When the counter matches the reference, the output is set to high and the counter continues counting. When the counter fills up, this signifies the end of the period and the output is again set to low and the process repeats. Clearly, the clock must be very fast to keep the switching frequency in the usual 20-kHz range. For an 8-bit system (256 counts) to have a 20-kHz switching frequency, the clock must run at 256 times 20 kHz, or 5.12 MHz. The remaining portions of the system are identical to the case where the input command signal is an analog voltage.

12.4.4 Signal Conditioning

Regardless of whether the amplifier is fed from an analog or a digital source, the issue of driving the output transistors remains. Two factors need to be considered: timing and isolation. The timing of the signals is important to assure consistent operation. The nominal PWM waveform may require slight timing adjustments. This is due to the fact that transistors will often turn on more quickly than they turn off. There is the danger that when switching transistors 2 and 3 on, 1 and 4 may take a finite time to turn off completely. It is often helpful to insert a short delay in the rising edge (turn-on) of the gate drive signal. This is easily accomplished with a one-shot and an AND gate. Sending the gate drive signal through a one-shot that triggers on the positive edge will produce a pulse that is low for a short time. This delay can be set by the one-shot's external RC time constant and is usually on the order of 1 μs or less. The output of the one-shot as well as a direct output are fed into an AND gate. This effectively delays the base drive on the positive transition.

The second consideration is isolation. Isolation of the drive signals and the output transistors is necessary because

- Switching the large amounts of current generates large amounts of noise that may be transmitted back into the logic circuits if no isolation were present.
- The voltage levels of the output transistors may differ greatly from the logic levels.

As an example, consider the MOSFET H-bridge shown in Figure 12.23. Say that the voltage V_{cc} is 50 V and the load can be regarded as 9 Ω. Consider the case where 1 and 4 are on, and let the effective resistance for the MOSFET in the fully on condition be 0.5 Ω. To keep M_4 in the fully-on state, the gate voltage needs to be significantly higher

V_{CC}

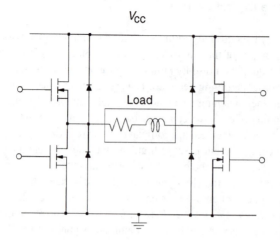

Load

Figure   12.23   N-Channel   MOSFET H-Bridge

than the source voltage. Presume that a CMOS logic signal of 10 V (relative to ground) is sufficient. M_1, however, is a different story. If M_1 is on, then the source of M_1 is at a voltage (relative to ground) of about 48 V to the voltage drop across the load and M_4. Clearly, there is a level shifting required and some method for generating a voltage that may be higher than the positive rail. (In the simpler case of an H-bridge using complementary devices, the requirement for voltages higher than the supply rail may not exist, but the level shifting issue remains.)

To deal with this, two methods of isolation are commonly used. The first is through the use of transformers. Transformers are inherently ac devices and have a frequency response that rolls off at high frequencies. Consequently, applying a sharp pulse at the input terminals results in a somewhat smoothed pulse at the output terminals. Moreover, as the duration of the pulse is extended, the amplitude of the output decays. To overcome this, modulation schemes can be used that encode the pulse information on a higher-frequency carrier. The energy of the modulated signal can be used to provide dc power, which is referenced to the source terminal of each MOSFET, and the signal can also be demodulated and the original information signal can be used to drive the gate in a dc fashion.

Another is the use of optical isolators (optoisolators). Unlike the transformer, the optoisolator is a dc device, so the signal will not decay for pulses of long duration. Since no power can be transmitted through optoisolators, they are better from an isolation viewpoint, but separate circuitry to develop a power source is required. Several *bootstrap* schemes that use a capacitor as a temporary energy source exist in databooks. These do have the limitation that when the duty cycle is held at an extreme for an extended period of time, the capacitor will eventually leak and the voltage will drop to an inadequate level. In this case a separate circuit for power generation for the gate drive is necessary.

12.5 POWER OP-AMPS

12.5.1 Overview

Until recently, the design of amplifiers from discrete components was the only option available. Within the last several years, *power op-amps* have been developed. These are functionally similar to a regular op-amp, but the voltage and currents to be delivered range up to 200 V and 30 A, respectively. Both linear and PWM designs are available in integrated-circuit form. The discussion here is directed at the linear variety, since integrated-circuit shortcuts for PWM amplifiers were discussed earlier.

12.5.2 Advantages and Disadvantages

The fundamental advantage of the power op-amp is that the user is relieved of all the design issues and much of the debugging. Power op-amps do require compensation to assure stability; in general, much more attention is required than for a regular op-amp. The key drawback is the cost. For a comparable level of performance, the component cost for a design consisting of discrete components is almost always lower than that of a power op-amp. However, the added simplicity and reduced assembly and debug time make a strong case for use of the prepackaged amplifier, particularly if only a small number of systems are built.

12.5.3 Application Hints

Compensation of any op-amp is based on the same principles. The essential notions of this procedure have been saved until now; even though this discussion occurs in conjunction with power op-amps, these techniques apply to any linear voltage amplifier. As a general rule, any amplifier can be stabilized, but the frequency response of the amplifier will be degraded accordingly.

12.6 ADDING ROBUSTNESS AND TROUBLESHOOTING

Although a power amplifier may behave well when tested after construction, it may still perform poorly in an actual application if precautions are not taken. A good design will account for the difficulties that may be attributable to the nature of the load to be driven, poor-quality power supply voltages, and problems resulting from excessive heat buildup.

12.6.1 Power Supply Considerations

The design of the power amplifier should take into account the nature of the power supply. It is important that the amplifier's current output be limited so that the power supply is not being asked to source excessive current. Further, any voltage spikes that may occur in the power amplifier (such as inductive kickback) should not be passed back into the power supply, as these would corrupt the steady voltage that should appear at the power rails of nearby circuits. Further, any spikes or hash that appear on the ac supply lines should not be passed through the power supply to the amplifier.

The main mechanism for decoupling the power supply from the amplifier is through decoupling capacitors (and occasionally, inductors). Since the power supply is normally connected to the amplifier through cables, the inductance in the cables can be significant when the amplifier output changes. To counteract this, capacitors are placed between the point where the cables enter the amplifier and ground. A coarse rule of thumb is to provide 10 µF of capacitance per ampere of peak output current. These capacitors will typically be of the electrolytic type (packaged in a small can) or tantalum (packaged in a dipped disk). These are often referred to as *low-frequency decoupling caps*. Although these are good at isolating larger but low-frequency transients, they are typically not good at blocking higher-frequency transients, since these relatively large capacitors have a tendency to pass higher frequencies. To block these higher frequencies, a smaller capacitor, typically 0.1 µF, is placed in parallel with the larger capacitor. In some instances it is helpful to add a small resistor in series with the small capacitor, on the order of 10 Ω or less, to add damping. Inductors in series with the supply cables are also sometimes used to block very high frequency noise, but these techniques are more common in high-frequency (on the order of MHz) amplifiers and are usually not necessary for the types of amplifiers used to drive mechanical devices.

The other mechanism of isolation is directed at large voltage spikes that may be passed from the ac line into the power supply. These may be clamped by using metal-oxide varistors (MOVs), which act like nonlinear resistors whose resistance decreases sharply at high voltages. These devices are connected across the ac line terminals and absorb any spikes that occur on the line. Since they dissipate the energy in the voltage spikes, extremely large voltage surges can destroy a MOV. As always, for construction of a small number of prototypes, generous overdesign is encouraged. A device that serves essentially the same function is the TransZorb, which acts like a pair of back-to-back zener diodes with a very fast turn-on time. A voltage spike across the ac line will bias one diode forward and one backward, which effectivly clamps the high-voltage transient.

12.6.2 Thermal Considerations

Thermal considerations have been mentioned previously, mainly in the context of thermal matching of bias networks and the V_{BE} voltages of the output transistors. The ideal solution is to minimize the temperature rise in the components through good mounting and heat sinking. For all but the most demanding applications, the heat generated by the devices is ultimately shed to the surrounding air. To carry the heat away most effectively, a good design will minimize the thermal resistance between the device and the heat sink. A large heat sink is used, and the thermal resistance between the heat sink and the surrounding air is minimized.

To minimize the thermal resistance between the device and the heat sink, good mounting practices should be followed. For some devices the metal case will be connected electrically to one of the output pins. In some instances (such as the case being grounded, or each device being mounted on individual electrically isolated heat sinks), this is not a concern and the device can be attached directly to the heat sink. In general, however, several devices with differing electrical case connections will share a heat sink. To prevent short circuits, mica or rubber insulating pads are used. These pads are placed between the device and the heat sink and the device is attached with small screws. In addition, insulating housings are used to isolate the screws electrically as well as the leads from the heat sink. When no pad or a mica pad is used, thermal grease should be applied to all mating surfaces. The soft rubber pads are designed to be used without grease. When insulating pads are used, particularly the softer rubber variety, care should be taken not to overtighten the screws, which may result in warping of the case and cracking of the silicon substrate.

Because of their high thermal conductivity, virtually all heat sinks are made out of aluminum. The surface on which the device is to be mounted should be flat and smooth. Ready-made extruded heat sinks are perfectly adequate in this regard. A lower-cost alternative for less-demanding situations is to use aluminum angle with a smooth finish. Another option is to mount the power transistors to the interior of the enclosure in which the amplifier is being built. Since these are often aluminum castings, machining the areas where devices are mounted is often necessary. Care should be taken to clean the surface thoroughly, since tiny metal chips or other debris will puncture the insulation, causing a short and contributing to warping of the case.

Finally, for demanding applications, free convection from the heat sink to the surrounding air may be inadequate. In such instances a small cooling fan should be positioned so that it forces air to flow over the heat sink to enhance the heat transfer. When a fan or fans are used, a good design will draw filtered air from the outside of the enclosure and create a positive pressure inside the enclosure. This will reduce the dust and contaminants that settle on the components.

12.7 PROBLEMS AND DISCUSSION TOPICS

1. Attach an existing power amplifier to a resistive load (if power resistors are not available, try a light bulb). Instrument it so that the current and voltage to the load can be measured and, if possible, the current and voltage to the amplifier. Measure the performance characteristics (command versus output) over the full range of operation. Look carefully at the behavior near zero for nonlinearity and dead zone.

2. If the amplifier in Problem 1 was a voltage amplifier, do the same with a current (transconductance) amplifier, or vice versa.

3. Instrument the amplifier(s) of Problems 1 and 2 with internal temperature measurements and record temperature at several steady-state operating points (record the transient to see how long it takes to reach steady-state operation).

4. Replace the resistor in Problem 1 with a motor of appropriate size for the amplifier being used. Use an oscilloscope or fast data acquisition system to find the *rise time* for the current, that is, how long the current takes to reach a specified fraction of its steady-state value. How does this electrical time constant compare to the motor's mechanical time constant (the time associated with the speed transient)? If possible, examine the transients in both voltage and current for both voltage and current amplifiers.

5. Figure 12.13 describes about the simplest possible linear amplifier. Following the data sheet for the particular (small) bipolar power transistors that you choose, build such an amplifier. Make sure to use the lowest-possible cost components — this process is instructive but sometimes destructive! Use an adequately sized heat sink (the datasheet will specify heat sink size).

(**a**) Connect a resistive load, and repeat the studies of Problem 1. Size the resistor so that it will not draw too much current at the maximum output voltage.

(**b**) Assuming that (a) is working (with power amplifiers, one never knows), study the frequency response of the system by applying sinusoidal command signals over a frequency range of dc to 100 kHz. This is a high enough range to cover most mechanical control problems.

(**c**) Examine the transient responses for various sizes of command signal step inputs.

(**d**) Replace the resistor with one that has a lower resistance value (but similar power-handling capability). Examine the amplifier behavior as the command increases beyond the nominal current output limit of the amplifier.

(**e**) Add an inductor in series with the (larger) resistor. Starting with very small step changes in the command, examine the transient behavior of the amplifier. If the amplifier does not have internal diode protection (Figure 12.11), add such protection and observe the behavior.

(**f**) Replace the resistor–inductor load with a motor and study the behavior. Include the frequency sweep test of part (b).

6. Repeat the studies of Problem 5 using the amplifier configuration with internal feedback (Figure 12.17).

7. Figure 12.22 shows a simple bipolar H-bridge configuration. The H-bridge is the most common way of configuring switching amplifiers to run motors in both directions. Build such an amplifier and build a logic input circuit so that there is a direction input and an on–off input, and examine its switching behavior when connected to load resistors of various sizes. Repeat with the resistor–inductor load, then with a motor. Use a signal generator to supply a square wave.

8. Build a logic-based PWM generator that accepts an analog input. Use it to operate the amplifier with a resistive load attached. Replace the resistor with a motor and observe the behavior at various duty cycles.

9. Examine the frequency response of the system of Problem 8 and compare the results with those of Problem 5f.

10. Repeat Problems 7, 8, and 9 using the MOSFET configuration of Figure 12.23.

Bibliography

ARTWICK, B. A., *Microcomputer Interfacing*, Prentice Hall, Englewood Cliffs, NJ, 1980.

BECKWITH, T. G., and R. D. MARAGONI, *Mechanical Measurements*, Addison-Wesley, Reading, MA, 1990.

CANDY, J. C., and G.C., TEMES, *Oversampling Delta–Sigma Data Converters: Theory, Design and Simulation*, IEEE Press, New York, 1992.

CROCHIERE, R. E., and L.R, RABINER, Interpolation and Decimation of Digital Signals—A Tutorial Review, *Proceedings of the IEEE*, Mar. 1981.

EGGEBRECHT, L. C., *Interfacing to the IBM Personal Computer*, Macmillan Computer Publishing, New York, 1990.

Electro-Craft, *DC Motors, Speed Controls, Servo Systems*, Electro-Craft Corporation, Hopkins, MN, 1980.

FREDERIKSEN, T. M. and J. E. SOLOMON, A High-Performance 3-Watt Monolithic Class B Power Amplifier, *IEEE Journal of Solid-State Circuits*, Vol. SC3, pp. 152–160, June 1968.

FRIEDMAN, S. B., *Logical Design of Automation Systems*, Prentice Hall, Englewood Cliffs, NJ, 1990.

GRAY, P. R., A 15-W Monolithic Power Operational Amplifier, *IEEE Journal of Solid-State Circuits*, Vol. SC7, pp. 474–480, Dec. 1972.

GRAY, P. R., and R. G., MEYER, *Analysis and Design of Analog Integrated Circuits*, 3rd ed., Wiley, New York, 1993.

GREENFIELD, J. D., *Practical Digital Design Using ICs*, Wiley New York, 1983.

Honeywell, *Hall Effect Transducers*, Micro Switch Division, Honeywell Corporation, Minneapolis, MN, 1982.

HOROWITZ, P. and W. HILL, *The Art of Electronics*, Cambridge University Press, Cambridge, 1989.

ILC, *Synchro Conversion Handbook*, ILC Data Device Corporation, Bohemia, NY, 1973.

INTEL, 16-*Bit Embedded Controllers*, handbook, part no. 270646-003, Intel Corporation, Santa Clara, CA, 1991.

JUNG, W. G., IC *Op-Amp Cookbook*, 3rd ed., Howard W. Sams, Indianapolis, IN, 1989.

KOHAVI, Z., *Switching and Finite Automata Theory*, McGraw-Hill, New York, 1970.

LANCASTER, D., *CMOS Cookbook*, 2nd ed., Howard W. Sams, Indianapolis, IN, 1989.

NACHTIGAL, C., *Instrumentation and Control*, Wiley, New York, 1990.

NASAR, S. A., *Handbook of Electric Machines*, McGraw Hill, New York, 1987.

NATIONAL SEMICONDUCTOR, *General Purpose Linear Devices Databook*, part no. 400026, National Semiconductor, Inc., Santa Clara, CA, 1989.

PEATMAN, J. B., *Design with Microcontrollers*, McGraw-Hill, New York, 1988.

RITTERMAN, S., *Computer Circuit Concepts*, McGraw-Hill, New York, 1986.

SANDIGE, S. S., *Modern Digital Design*, McGraw-Hill, New York, 1990.

SCHAMUMANN, R., Design of Continuous-Time Fully Integrated Filters, *IEEE Proceedings*, Vol. 136, No. 4, Aug, 1989.

SEDRA, A. S., and K. C. SMITH, *Microelectronic Circuits*, Saunders College Publishing, Philadelphia, 1991.

SENSYM, *Solid-State Sensor Handbook*, SenSym, Inc., Sunnyvale, CA, 1989.

Stone, H. S., *Microcomputer Interfacing*, Addison-Wesley, Reading, MA, 1982.

TEXAS INSTRUMENTS, *Optoelectronics and Image Sensors Data Book*, Texas Instruments Inc., Dallas, 1990.

Wakerly, J. F., *Microcomputer Architecture and Programming*, Wiley, New York, 1989.

Index

A

A/D, *see* Analog-to-digital conversion
Absolute encoder, 35, 188
Absolute value circuit, 201
Ac induction motor, 132
Ac motors, 2
Actuators, 1
ADC, *see* Analog-to-digital conversion
Aliasing, 156, 157
Analog computer, 198
Analog, 2
Analog-to-digital conversion, 139
Antialiasing filter, 158
Arrival condition, 90
Asynchronous circuits, 59
Asynchronous inputs, 44

B

Back-emf, 111, 117, 125, 130, 183, 187
Bandwidth, 210
BCD, *see* Binary-coded decimal
Bi-directional parallel port, 86
Bifilar winding, 98, 99
Binary, 8
Binary-coded decimal, 188, 189
Bipolar junction transistor, 219
Boolean algebra, 9
Boole, George, 9
Boxcar filter, 154
Breakdown voltage, 200
Brushless motor, 3, 132
Bubble memory, 84

Buffer zone, 8
Bus, 25, 85
Bus master, 77
Bus protocol, 77

C

Cascade control, 128
Central processing unit (CPU), 83
CISC, *see* Complex instruction set computers
Clamping circuit, 201
Class-A amplifier, 223
Class A-B amplifier, 225, 228
Class-B amplifier, 223
Clock, 44
CMOS, 22, 83
Combinational logic, 13
Commutation, 114
Commutator, 3, 114
Comparator, 203
Complementary metal oxide semiconductor (CMOS), 22
Complex decision making, 2
Complex instruction set computers (CISC), 89
Content-addressable memory, 87
Control signals, 76
CPU, *see* Central processing unit
Critical race, 65
Crossover distortion, 224
Cruise control, 5
Current amplifier, 214
Current control, 124
Current-controlled amplifier, 124

T

U-V

W, X, Y, Z

W, X, Y, Z

T

U-V